ELECTROLYTIC IN-PROCESS DRESSING (ELID) TECHNOLOGIES

Fundamentals and Applications

ELECTROLYTIC IN-PROCESS DRESSING (ELID) TECHNOLOGIES

Fundamentals and Applications

Edited by

Hitoshi Ohmori
Ioan D. Marinescu
Kazutoshi Katahira

CRC Press
Taylor & Francis Group
Boca Raton London New York

CRC Press is an imprint of the
Taylor & Francis Group, an **informa** business

CRC Press
Taylor & Francis Group
6000 Broken Sound Parkway NW, Suite 300
Boca Raton, FL 33487-2742

First issued in paperback 2017

© 2011 by Taylor & Francis Group, LLC
CRC Press is an imprint of Taylor & Francis Group, an Informa business

No claim to original U.S. Government works

ISBN-13: 978-1-4398-0036-2 (hbk)
ISBN-13: 978-1-138-07402-6 (pbk)

Visit the Taylor & Francis Web site at
http://www.taylorandfrancis.com

and the CRC Press Web site at
http://www.crcpress.com

Contents

SECTION I   *Fundamentals of the ELID Process*

SECTION II   *ELID Operations*

SECTION III   *Case Studies*

Introduction

The needs of manufacturing and technological industries for materials that have certain characteristics have driven the development of hard and ultrahard materials. These characteristics include the ability to retain strength at increasingly higher temperatures, to survive in any chemical environment, and to meet certain electrical or magnetic properties.

The market for hard and ultrahard materials is increasing at a very rapid rate. These materials are increasingly being used in aerospace and automotive industries because they are lightweight, offer high-temperature strength, have a high resistance to wear and corrosion, and need less lubrication. Because of the high hardness of these materials, it is not possible to effectively use other material removal methods commonly utilized with metals such as turning, milling, and drilling. Grinding with diamond wheels has become one of the primary methods used in machining of hard and ultrahard materials. At the same time, these materials are brittle, so grinding consists of a combination of microbrittle fracture and quasiplastic cutting. The mechanism of quasiplastic cutting, typically referred to as ductile-mode grinding, results in grooves on the surface that are relatively smooth in appearance. The transition from brittle to ductile-mode grinding is hard to predict because there is no universal mechanism for the grinding of brittle material. The material removal mechanism varies from one material to another, and it is hard to tell how many grinding grits will be in contact with the workpiece.

The problem commonly associated with grinding of hard and ultrahard materials is the surface defects induced during the grinding process. The fracture damages can be avoided or reduced by the careful choice of grinding parameters and control of the process to obtain ductile-mode grinding. But ductile-mode grinding is a costly process and requires a lot more energy than the brittle grinding, which in turn generates a high grinding temperature.

The use of ultrahard materials is limited in the industry because of high costs involved in the machining of these materials. The cost of grinding may be up to 75% of a component cost for these parts. Therefore, it is necessary to optimize the grinding process by any means. However, it is a very complicated task. A high material removal rate can be achieved by using cast-iron bonded diamond grinding wheels. These wheels are manufactured by mixing diamond abrasive grains, cast-iron powder or fibers, and a small amount of carbonyl iron powder, compacting it to the desired form under the pressure of 6 to 8 ton/cm^2, and then sintering it in an atmosphere of ammonia. Cast-iron bonded diamond wheels are not suitable for continuous grinding for a long period of time because of their poor dressing ability and frequent wear of the abrasive.

The electrolytic in-process dressing (ELID) technique has been introduced to conduct high-efficiency grinding of ceramics using cast-iron bonded wheels. ELID grinding was first proposed by the Japanese researchers Ohmori and Nakagawa in 1990. ELID grinding is based on the principle that the metal-bonded grinding wheel is dressed because of the electrolytic action between the grinding wheel and the fixed electrode. The ELID system essentially consists of a metal-bonded grinding wheel, electrolytic power source, electrode, and electrolytic coolant. The grinding wheel is connected to the positive terminal (which is the anode) and the fixed copper electrode is connected to negative terminal (which is the cathode) of the power supply. The wheel is dressed by removing a small amount of material with the help of electrolysis that occurs between the wheel and the electrode when direct current is passed through a suitable grinding fluid that acts as an electrolyte.

The grinding process can be made more economically efficient with online process control. This can be achieved by means of advanced sensing technology based on acoustic emission (AE). Acoustic emission has been in use since the late 1960s as a nondestructive testing method for the evaluation of fatigue and fracture of solid materials, and for the monitoring of flaw and crack growth in pressure vessels. Only two decades have passed since acoustic emission has been applied to the field of metal cutting and used as a tool to monitor machining operations.

Acoustic emission is referred to as the release of a transient elastic wave in the lattice of crystalline materials due to rearrangements of the internal structure of a material under deformation. The transient elastic wave is detected by the sensor attached to the material, and the sensor converts it into electric signals and sends them to the hardware system for further processing. As an acoustic emission sensor converts the mechanical energy carried by elastic waves into an electric signal, and the sensor is termed a transducer. The transducer most often used in acoustic emission applications is the piezoelectric transducer. A piezoelectric crystal is strained and an electric signal is generated due to the displacement of the surface to which it is attached. Acoustic emission ELID investigation is part of a new strategy for monitoring the process from a quantitative viewpoint.

The *Electrolytic In-Process Dressing (ELID): Fundamentals and Applications* is the first English overview of ELID processes and applications. The book contains three parts: Section I presents the fundamentals of the ELID process with correlations between the main parameters, Section II describes ELID operations, and Section III consists of case studies.

Section I contains Chapter 1 and Chapter 2. Chapter 1 delineates the fundamental process of ELID as well as the chemistry and physics of the process. Hardware required by the ELID process is also presented. Chapter 2 presents the types of ELID that are mainly applied to different configurations of grinding.

Section II contains Chapter 3 through Chapter 6, which are dedicated to ELID operations. Chapter 3 presents ELID grinding methods that are the most used ELID operations in industry. Chapter 4 is focused on the ELID lapping/grinding process, which is mainly the replacement of lapping and polishing (loose abrasive processes) with ELID lapping/grinding, which is a bonded process, but one that uses lapping/polishing kinematics. ELID honing is presented in Chapter 5 as well as different ways of applying ELID to the honing process. Chapter 6 presents an original method of ELID grinding of free-form surfaces using an original design.

Section III, which is composed of real case studies, will provide the most interest for industry. Chapter 7 presents nano-ultraprecision ELID and the latest developments in ELID nanogrinding, which includes applications to glass ceramic mirrors, small lens, and large-scale optics. Together with these applications, the equipment required is described, along with new designs. Chapter 8 is dedicated to a new concept of the micro-workshop, where all the machines tools and measurement devices are tabletop machines with high accuracy. Chapter 9 presents successful applications of ELID technology: ELID in the semiconductor industry (ELID grinding of Si wafers), in the mold and die industry, and in the microtools industry; plus the application of ELID in the optics industry is presented as a combination of the ELID grinding and MRF (magneto-reological finishing), which is a new method. The last part of Chapter 9 presents surface modifications as a future method for obtaining complex modifications of surfaces by using ELID in combination with other methods.

This book presents for the first time a new technology with various applications and the science behind this method. In order to improve any process, we must first understand the fundamentals of the process (physics, chemistry, and mechanics). Therefore, we have put together a book that we hope will anticipate many of the the questions that will arise from the investigation of ELID methods and applications.

Editors

Hitoshi Ohmori is a chief scientist in the Materials Fabrication Laboratory at RIKEN, Japan. He has a doctorate from the Graduate School of Engineering at Tokyo University. Ohmori is a member of the Japan Society of Precision Engineering (JSPE), Japan Society of Abrasive Technology (JSAT), Japan Society of Mechanical Engineering (JSME), and International Academy for Precision Engineering (CIRP). Ohmori's areas of research interest include ultraprecision, ultrafine, nanoprecision, and ultrasmooth machining processes required for the fabrication of advanced functional devices such as optical and electronic components.

Ioan D. Marinescu, PhD, is a professor in the Department of Mechanical, Industrial, and Manufacturing Engineering at the University of Toledo, Ohio. He is the director of the Precision Micro-Machining Center (PMMC). He has a doctorate from the Department of Manufacturing Engineering at the University of Galatzi, Romania. Marinescu is a member of the American Society of Mechanical Engineers (ASME), Society of Manufacturing Engineers (SME), American Ceramic Society (ACerS), and American Society for Abrasive Technology (ASAT). Marinescu's areas of research interest include tribological investigation of abrasive processes, minimum lubrication in metal cutting, investigation of friction and wear mechanisms, and molecular dynamics simulation of friction and wear.

Kazutoshi Katahira is a senior scientist in the Materials Fabrication Laboratory at RIKEN, Japan. He has a doctorate from the Graduate School of Science and Engineering at Ibaraki University, Japan. Katahira is a member of the Japan Society of Precision Engineering (JSPE), Japan Society of Abrasive Technology (JSAT), Japan Society of Mechanical Engineering (JSME), and International Academy for Precision Engineering (CIRP). Katahira's areas of research interest include nanomachining process management, analysis of nanoscale surface generation, nanolevel controlling on optimum tool condition, and surface functional modification techniques.

Contributors

Nobuhide Itoh
Ibaraki University

Hiroshi Kasuga
RIKEN

Jun Komotori
Keio University

Weimin Lin
Akita Prefectural University
Akita, Japan

Takashi Matsuzawa
Ikegami Mold Engineering Company

Masayoshi Mizutani
RIKEN

Tetsuya Naruse
Koriyama Technical Academy

Yoshihiro Uehara
RIKEN

Yutaka Watanabe
RIKEN

Shaohui Yin
Hunan University

Section I

Fundamentals of the ELID Process

1 Fundamentals of ELID

Hitoshi Ohmori, Ioan D. Marinescu,
and Kazutoshi Katahira

CONTENTS

1.1 DEVELOPED PRINCIPLE OF ELID

The electrolytic in-process dressing (ELID) grinding method is not a complex machining process but a grinding technique whose ability is maintained by assistance of the special ELID. One of the authors did not give the *in-process* term the meaning "continuous" because the electrolytic phenomenon did not seem to be so continuous but adaptive to the dressing requisition. Figure 1.1 shows the principle of the proposed ELID grinding method.[1–7] The wheel is at the +Ve pole with a brush smoothly contacted, and the electrode fixed below is at the –Ve pole. In the small clearance of approximately 0.1 mm between the –Ve and the +Ve poles, electrolysis occurs upon the supply of grinding fluid and an electrical current. The electrolysis is greatly helpful for maintaining of the protrudent grains on the wheel surface and chip removal.

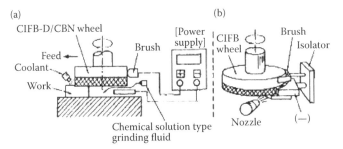

FIGURE 1.1   Schematic of ELI-grinding. (a) System construction. (b) Electrode detail.

1.2   ELID COMPONENTS

The essential elements of the ELID system are (a) a metal-bonded grinding wheel, (b) an electric power source, and (c) an electrolytic coolant (Figure 1.2). The most important feature is that there is no requirement for a special machine. The selection of these elements determines the properties realized by ELID.

Selection of the electrode for ELID depends on its material and number. As its material, a conductive material can be used (copper, stainless steel, carbon, etc.). The surface area opposite the grinding wheel is around one-sixth the surface area of the grinding wheel. Electrodes that generate dynamic pressure are best with large grinding wheels. Furthermore, tape electrodes, devices for moving electrodes, and removable electrodes are also available.

From the first stage of developing of ELID grinding, one of the authors utilized micrograin cast-iron bonded wheels. In previous research other metal bonded superabrasive wheels were used, such as nickel (electroplated), bronze, and cobalt bonded wheels. The different characteristics on ELID could be recognized. In use of nickel bonded wheels, almost the same mirror surfaces could be realized as cast-iron

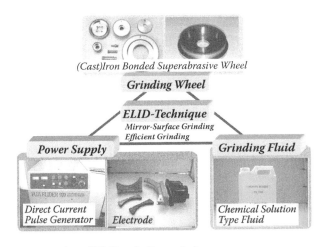

FIGURE 1.2   Construction of ELID grinding technique.

bonded wheels, but bronze bonded wheels did not easily show good adaptability to mirror surface grindings. These are from different "electrolytic dressabilities," that is to say, from controllability of electrolysis. Primarily, a cast iron or cobalt bonding material is used with a diamond grinding wheel. With copper grinding wheels, insulating oxide layers are difficult to produce. Furthermore, grinding wheels with a composite bonding material (metal + resin) are also used.

Regarding the power supply, a rippled current that was made by combination of a high-frequency pulse direct current (DC) and a continuous DC was revealed to be superior to the others for ELID mirror surface grindings. This is considered to be due to balancing of the electrolysis and the isolation of the wheel's bond for avoiding excessive electrolyzing, and contact between the wheel bond and work surface. Generally, a high-frequency (2–4 μs is standard) DC rectangular wave is preferable. And, with ordinary direct current or alternating current, uniform ELID is not possible.

A water-soluble grinding fluid (a dilution ratio of around 5%) with weak conductivity is used as the electrolyte (a general coolant that washes chips away) to avoid excess electrolyte elution. The grinding fluid should have the appropriate alkalinity (around pH 11).

1.3   ELID EFFECTS

The most important and significant effect of ELID is the realization of mirror surface quality grinding. Figure 1.3 shows surface qualities produced by cast-iron bonded diamond wheels. With a #1200 wheel of average grain size of 12 microns, the ground surface is made by brittle fracture mode removal. However, with the assistance of ELID, the ground surface removal is transferred to ductile mode by obtaining micrograin protrusions with a #4000 wheel of average grain size of 4 microns. Figure 1.4 shows the relationship between the wheel mesh size and finished surface roughness Rz. The second significant effect of ELID is stabilization

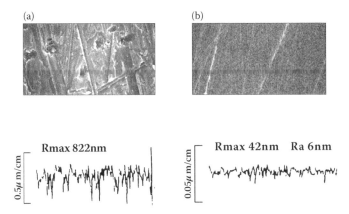

FIGURE 1.3   Ground surface quality. (a) Brittle mode grinding (#1200). (b) Ductile mode grinding (#4000).

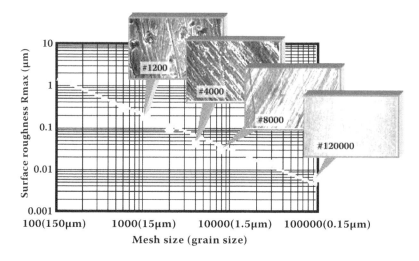

Wheel mesh size (#)	325	400	600	800	1000	1500	2000	4000	8000
Average grit size (μm)	44	37	30	20	15	10	8	3	2

FIGURE 1.4   Relationships between grain size and surface roughness.

of finish grinding forces. Figure 1.5 shows an example of grinding force change in the grinding of optical glass stabilized by ELID. Together with the realization of mirror surface quality grinding and quite lower grinding forces, stability of these properties should be mentioned. A mirror surface such as one in Figure 1.3b can be

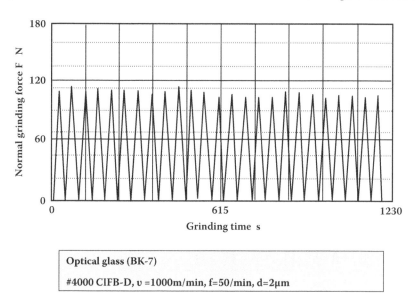

Optical glass (BK-7)

#4000 CIFB-D, υ =1000m/min, f=50/min, d=2μm

FIGURE 1.5   Stability of grinding force.

maintained for an extremely long time during the grinding operation and also has very good repeatability. This feature can make the technique applicable to practical uses.

1.4 ELID MECHANISM

Figure 1.6 shows the electrical behavior during ELID. At the time predressing starts (point 1), the surface of the trued wheel has good electroconductivity. Therefore, the current is as high as has been set on the power source, and the voltage between the +Ve pole and –Ve pole (wheel and electrode) is low. For several minutes, the bond material (mainly iron [Fe]) is removed by electrolysis. It is mostly ionized into Fe^{2+}. The ionized Fe forms hydroxides, Fe $(OH)_2$ or Fe $(OH)_3$. The ionized Fe reacts to form $Fe(OH)_2$ and $Fe(OH)_3$ according to the following equations:

$$Fe \rightarrow Fe^{+2} + 2e^- \tag{1.1}$$

$$Fe^{+2} \rightarrow Fe^{+3} + e^- \tag{1.2}$$

$$H_2O \rightarrow H^+ + OH^- \tag{1.3}$$

$$Fe^{+2} + 2OH^- \rightarrow Fe(OH)_2 \tag{1.4}$$

$$Fe^{+3} + 3OH^- \rightarrow Fe(OH)_3 \tag{1.5}$$

Next, these substances change into oxide substances such as Fe_2O_3. After these reactions occur, the electroconductivity of the wheel surface is reduced with the growth of the insulating substances. Thereby, the current decreases, and the working voltage becomes as high as that that has been set as the open voltage (denoted by 2 in Figure 1.6).

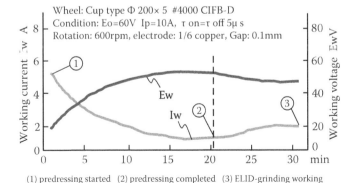

Wheel: Cup type Φ 200× 5 #4000 CIFB-D
Condition: Eo=60V Ip=10A, τ on=τ off 5μ s
Rotation: 600rpm, electrode: 1/6 copper, Gap: 0.1mm

(1) predressing started   (2) predressing completed   (3) ELID-grinding working

FIGURE 1.6 Electrical behavior of ELID. (1) Predressing started. (2) Predressing completed. (3) ELID grinding working.

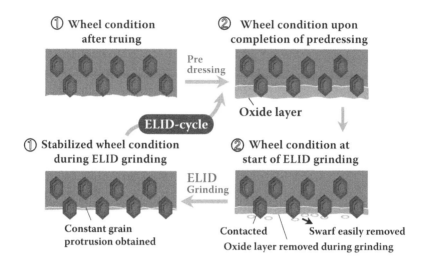

FIGURE 1.7   Mechanism of ELID grinding.

Figure 1.7 shows a schematic illustration of the mechanism of the ELID grinding process. After grinding with the predressed wheel begins, the protruding grains grind the workpiece. Accordingly, as the grains wear down, the oxide layer also becomes worn. The wear of the oxide layer also becomes worn. The wear of the oxide layer causes an increase in the electroconductivity of the wheel surface. Thus, electrolysis increases and the oxide layer can be recovered.

The mechanism of ELID grinding shown in Figure 1.7 is explained in detail as follows. First, the wheel bond is electrolyzed, which causes the abrasives to protrude appropriately (step 1). At the same time, this process produces nonconductive oxidized iron, which accumulates to form a layer of coating on the wheel surface, automatically reducing the electrolysis current, at which point the initial dressing is complete (step 2). When the actual grinding work is started in this state, the nonconductive oxide layer on the wheel surface comes into contact with the surface of the workpiece and is removed by friction. As this takes places, the abrasives start to grind the workpiece and consequently the wheel begins to wear (step 3). This reduces the insulation of the wheel surface, allowing the electrolytic current to flow again. As a result, the entire process starts again with the electrolysis of the wheel bond where the nonconductive oxide coating between the worn out abrasives has become thin (step 4), allowing the abrasives to protrude again (process returns to step 2). The protrusion of the grains therefore remains constant in a general sense. This is the ELID cycle. The cycle described can be changed according to the grain size of the employed wheel, the electrolytic conditions, the workpiece, and the grinding conditions, which can be optimized for practical performance. The bonding material can be removed electrically, without removing anything else, and regardless of the size of the abrasives, making this technique applicable with superabrasive grinding wheels. ELID works most effectively under the balance of the thickness of the oxide layer and the depth of the etched layer on the wheel surface. Using electrolysis

FIGURE 1.8 Microscopic photographs of the wheel surface after grinding with ELID (left) and without ELID (right).

prevents excess dressing caused by hydroxide layers and oxide layers (both of which are insulators) that form on the surface of the grinding wheel, and this significantly extends the service life of the grinding wheel (by a factor of 10 or more over conventional grinding).

Figure 1.8 compares microscopic photographs of the wheel surface after grinding with ELID (left) and without ELID (right) to demonstrate the effectiveness of ELID. As shown in the figure, for the grinding without ELID, the abrasives are worn out, scratch marks caused by the workpiece are observed on the bond surface, and the sharpness is poor; whereas for the grinding with ELID, the bond surface maintains abrasive protrusion by electrolysis.

1.5 EFFICIENT GRINDING WITH ELID

When ELID is applied to rough grinding, lower grinding forces can be maintained by protruding coarse abrasives. In ELID rough grinding, the rate of electrolysis is higher than the generation rate of the insulating layer of the wheel bond. That is because the abrasive wears down and the coarse chips remove the insulating oxide layer on the wheel rapidly. Figure 1.9 shows the difference on the normal grinding force of Si_3N_4 (silicon nitride) ceramics using a #170 wheel when ordinary and ELID grinding are performed. The latter is approximately three times lower than the former. Up to 6000 mm^3/min could be achieved on a conventional grinder with ELID. Figure 1.10 shows examples of shape grinding with ELID for Si_3N_4 ceramics.

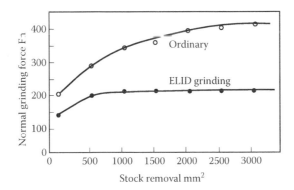

FIGURE 1.9 Grinding force for ceramics.

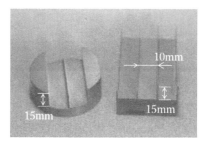

FIGURE 1.10   Efficient shape-grinding of ceramics.

According to the results, stability of efficient grinding processes can also be ensured easily by ELID. The authors expect ELID grinding techniques to be used widely from rough grindings to mirror finish processes in manufacturing.

1.6   ELID GRINDING WITH SUPERFINE ABRASIVES

Ultrasmooth ELID machining is a method aimed at realizing an ultrasmooth surface using ELID grinding. Specifically, the aim is to produce a surface roughness at the nanolevel. This requires ensuring sufficient machining action of the grinding wheel abrasives as well as stability. Nanometric finishing with ELID was devised on the constant pressure grinding. A workpiece was softly pressed on the grinding wheel surface under ELID. The schematic illustration of this system is shown in Figure 1.11.[8] SiC (silicon carbide) ceramic was finished smoothly down to about 5 nm in peak to valley (PV) as shown in Figure 1.12 using #3,000,000.

1.7   TRUING FOR ELID GRINDING

Before beginning the grinding process, the grinding wheel is trued and dressed. This is required in order to reduce the eccentricity of the wheel resulting from mounting the wheel on the spindle. Truing is also carried out to create a wheel having a desired shape or to correct a dulled profile. Truing of conventional wheels is easily performed with a diamond dresser. Although metal-bonded superabrasive

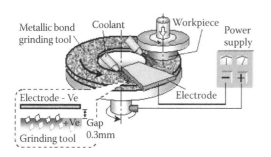
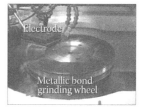

FIGURE 1.11   Nanoprecision surface finishing.

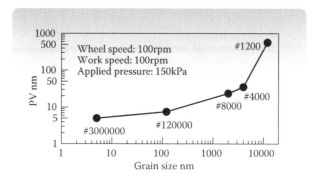

FIGURE 1.12 Relation between grain size and roughness on SiC.

wheels have many good features, such as higher rigidity, they are difficult to true. Several studies have examined areas of truing and dressing. The majority of these investigations examined the effect of diamond dressing on surface roughness.[9–11] A number of studies have focused on the mechanics of dressing[12] and the development of a mathematical model to characterize the topography of the wheel.[13] In one investigation, the present authors measured the protrusion height of abrasive grains after truing and dressing.[14] However, an efficient method for the truing of metal-bonded superabrasive wheels has not yet been reported. Metal-bonded diamond grinding wheels can be efficiently trued using a newly developed method known as electrodischarge (ED) truing, the concept of which was first introduced by the present authors.[15] Electrodischarge truing offers the following advantages: (a) applicability to all metal-bonded wheels and electrically conductive resinoid-bonded wheels, (b) precise truing due to on-machine application, (c) applicability to small wheels and thin wheels due to a small applied force, (d) high efficiency, and (e) ability to produce various profiles or shapes. These advantages have resulted in the application of ED truing in a number of studies.[16–17]

To true or profile metallic bond grinding wheels for desktop ELID grinding, microplasma discharge truing process and its specific truing device were developed. In this system, the grinding wheel is rotated comparatively slow, while the cathode is turned and moved toward and along the axis of the grinding wheel. The specific power supply is commonly used for truing and ELID. The principle of this method is shown in Figures 1.13 and 1.14. Figure 1.15 shows a view of a truing device. Copper tungsten was used for an electrode material. A small amount of pressurized mist fluid is supplied between the wheel and the electrode. During this truing operation, the bonded material is evaporated into plasma, and is quickly removed. Figure 1.16 shows a sharply trued wheel edge. Figure 1.17 shows the relationship between the truing conditions and the decrease in the grinding wheel diameter. The rough truing efficiency was 0.2 mm for a 15-minute period of material removal in the wheel diameter and the fine truing efficiency was approximately half of that (ED truing was applied on a bronze-and-iron bonded hybrid wheel).

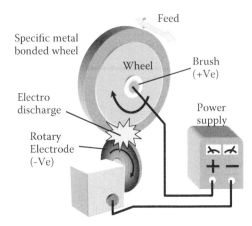

FIGURE 1.13   Principle of microplasma discharge truing.

FIGURE 1.14   Overview of machine setup with electrodischarge truer.

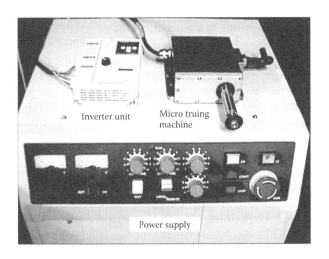

FIGURE 1.15   View of microplasma truing devise.

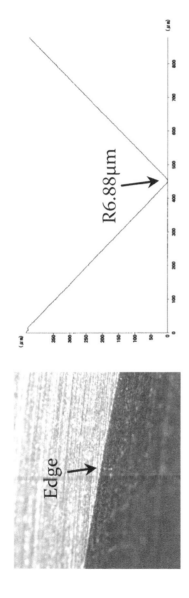

FIGURE 1.16   Sharply trued grinding wheel edge.

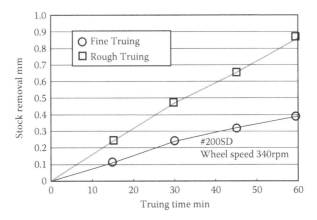

FIGURE 1.17   Change in stock removal in wheel diameter with truing time.

1.8   ELID FLUIDS: INVESTIGATION OF BOND OXIDATION KINETICS IN THE ELID PROCESS

1.8.1   INTRODUCTION

ELID grinding is a process that employs a metal-bond superabrasive wheel with an in-process dressing through electrolytic actions. The electrolytic process continuously exposes new sharp abrasive grains by dissolving the metallic bond around the superabrasive grains in order to maintain a high material-removal rate and to obtain a consistent surface roughness.[18] The basic system of ELID consists of a metal-bond grinding wheel, a power supply, and a type of alkaline coolant that usually has a high electrolytic property. There are two electrodes in this system. The metal-bonded wheel is applied as the positive electrode, while a copper brush is applied as the negative electrode. The power supply provides electric power to these two electrodes. The gap distance between two electrodes is usually 0.1 to 0.3 mm.[18]

On one hand, the application of ELID makes the in-process dressing possible, so the dressing time is eliminated and the dressing cost is reduced. On the other hand, the oxide layer generated on the grinding wheel surface significantly reduces the normal grinding force compared to conventional grinding, so a high-quality ground surface can be achieved.[19,20]

1.8.2   EXPERIMENTAL SYSTEM

In the fundamental experiment studies, a metal-bond sample without abrasives is used in an ELID cell to simulate the oxide layer generating process. Similar to the grinding process, the metal-bond sample is made the anode, while the copper electrode is made the cathode. The electric parameters, such as voltages of power supply, the gap between two electrodes and the velocity of coolant flow between two electrodes, are adjusted. Figure 1.18 shows the basic principle of the fundamental ELID experimental system.

Figures 1.19 and 1.20 show the experimental system. Figure 1.21 shows a sample of metal bond wheel. There are two tanks and two pumps in this system. The coolant

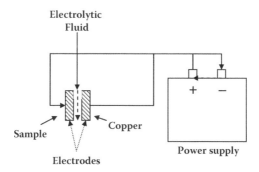

FIGURE 1.18 Principle of ELID system.

is restored in tank A. Pump A withdraws coolant from tank A and sends it into the gap between the two electrodes. The power supply provides a set voltage to the electrodes. After going through the reaction between two electrodes, the coolant falls into tank B. Then pump B withdraws the coolant from tank B and transfers it back into tank A. Therefore, the coolant is circulated in this system. The metal-bond sample contains two kinds of metal, copper (Cu) and tin (Sn), which can be oxidized. The copper electrode shares the same dimension as the metal-bond sample. The parameters we change are the voltage between two electrodes, the distance between electrodes, and the velocity of coolant flow. A 3^3 factorial experiment is conducted.

Voltage: 55 V, 70 V, 90 V
Gap: 0.3 mm, 0.5 mm, 0.7 mm
Velocity of coolant flow: 57.6 ml/min, 216 ml/min, 480 ml/min

There is an oxide layer generated on the electrolyzed surface of the metal-bond sample. The thickness of the metal-bond sample is measured before and after every experiment under combinations of different parameter levels. Then the influence of three parameters on oxidation kinetics can be obtained.

1.8.3 EXPERIMENTAL RESULTS

1.8.3.1 Electrical Behavior

At the initial stage of the electrolytic process, there is no oxide layer on the surface of the sample, so its conductivity is high. The potential between two electrodes is low while the current is high. The rate of generation of the Sn and Cu oxide layer is the highest during this stage, so the rate of change of voltage and current is the highest. As the oxide layer is generated on the sample surface continuously, the resistance of the sample increases and the conductivity decreases. From Ohm's law, the potential between two electrodes increases as the resistance increases. The current also decreases because of the increasing resistance. This causes the generation speed of the oxide layer to decrease, so the rate of change of voltage and current gradually becomes stable.

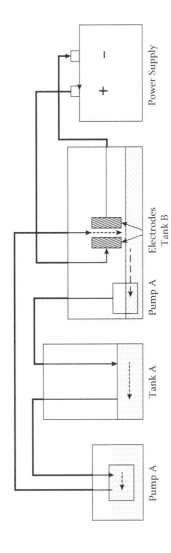

FIGURE 1.19 Experimental setup system.

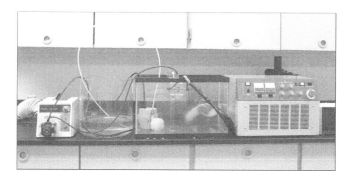

FIGURE 1.20 Experimental system.

When the voltage is 55 V, the current becomes stable at the 25-minute mark under gaps of 0.3 mm (Figure 1.22), whereas reaching stability takes 30 minutes under 0.5 mm (Figure 1.23). The decreasing slope of current under the distance of 0.3 mm is steeper than 0.5 mm. When the gap is fixed at 0.3 mm and the voltage is raised to 70 V, the current becomes stable at the 20-minute mark (Figure 1.24), which is earlier than under 55 V. Then the velocity of flow is reduced to 216 ml/min (Figure 1.25). The current becomes stable at the 30-minute mark, which takes longer than 25 minutes at velocity of flow 480 ml/min.

1.8.3.2 Oxide Layer Thickness

Figures 1.26 through 1.28 show the influences of different voltages on thickness change are compared. At every voltage level, the experiments was repeated eight times, with 5 minutes, 10 minutes, 15 minutes, 20 minutes, 25 minutes, 30 minutes, 35 minutes, and a maximum time of 40 minutes, respectively. The thickness changes of the sample were measured during every experiment.

All of the following plots are obtained at the velocity of flow 480 ml/min. Figure 1.26 shows that the growth rate of the oxide layer is higher at 90 V than at 70 V. Figure 1.27 shows that the growth rate of oxide layer is higher at 0.2 mm than at 0.3 mm. Figure 1.28 shows that the growth rate of the oxide layer is higher at velocity of flow 480 ml/min than at velocity of flow 216 ml/min.

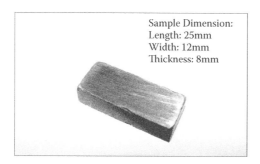

Sample Dimension:
Length: 25mm
Width: 12mm
Thickness: 8mm

FIGURE 1.21 Metal-bond sample.

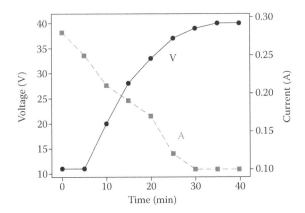

FIGURE 1.22 Electrical behavior voltage: 55 V; gap: 0.5 mm; velocity of flow: 480 ml/min; coolant: 9106 5%.

1.8.4 Discussion

From the experimental results, it can be concluded that the three effects influence the change rate of the current intensity and the oxide layer generation rate. Then we can conclude that the change of current intensity will influence the oxide layer generation rate. From the knowledge of fundamental electrochemistry, the total current density is the sum of the partial currents due to transportation of each type of ion:

$$i_j = FEz_jc_ju_j = F\frac{U}{D}z_jc_ju_j \tag{1.6}$$

The ion type is marked by j, z_j is the charge number of ions j, and F is the Faraday constant. The voltage between the two electrodes is U, and the distance is D. c_j is

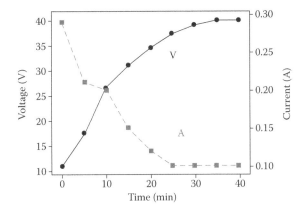

FIGURE 1.23 Electrical behavior voltage: 55 V; gap: 0.3 mm; velocity of flow: 480 ml/min; coolant: 9106 5%.

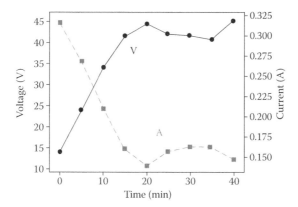

FIGURE 1.24 Electrical behavior voltage: 70 V; gap: 0.3 mm; velocity of flow: 480 ml/min; coolant: 9106 5%.

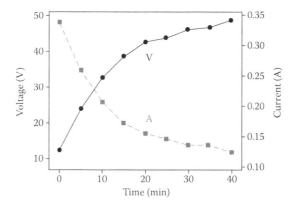

FIGURE 1.25 Electrical behavior voltage: 70 V; gap: 0.3 mm; velocity of flow: 216 ml/min; coolant: 9106 5%.

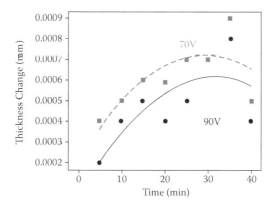

FIGURE 1.26 Thickness change 70 V versus 90 V.

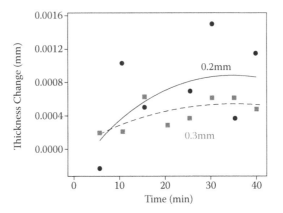

FIGURE 1.27   Thickness change 0.2 mm versus 0.3 mm.

ion volume concentration and u_j is the mobility of the ions.[21] If the same material is contained in the metal-bond sample, z_j is the same. Both c_j and u_j can be affected by the voltage, gap, and velocity of flow between the two electrodes.

We can conjecture the following equation:

$$T = e^c U^\alpha D^\beta S^\gamma + \varepsilon \tag{1.7}$$

U is voltage, D is gap, and S is velocity of flow. c, α, β, and γ are exponents we need to find. This equation implies that the thickness change rate with three parameters may not be a "straight line." If the change rate is a straight line, α, β, and γ would be 1.

A 3^3 factorial design is conducted and the response is the thickness of the oxide layer. The method of least squares is used, and then a fitted regression model can be obtained.

$$T = e^{-15.1299} U^{2.5459} D^{(-0.0555)} S^{0.492994} \tag{1.8}$$

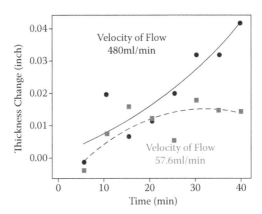

FIGURE 1.28   Thickness change velocity of flow 480 ml/min versus velocity of flow 216 ml/min.

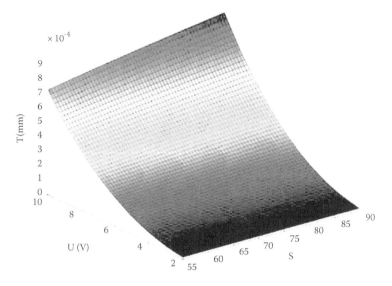

FIGURE 1.29   The response surface gap: 0.3 mm.

The thickness change response surface is plotted. Figures 1.29 and 1.30 clearly show the trend that the thickness change increases with the voltage and velocity of flow, respectively.

The positive relationship between thickness change and velocity of flow is more evident in Figures 1.31 and 1.32. However, the thickness change with the change of distance is too small to be detected. With fixed voltage and velocity of flow, some

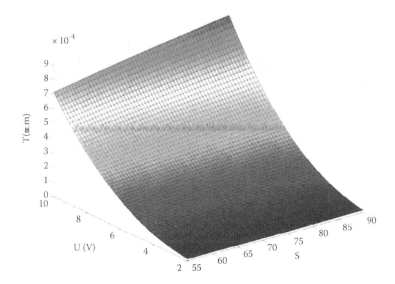

FIGURE 1.30   The response surface gap: 0.5 mm.

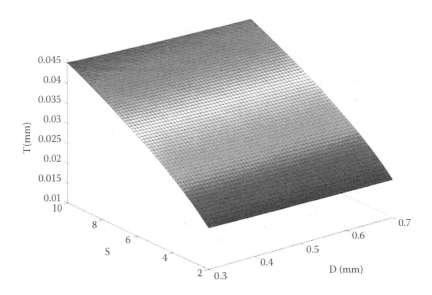

FIGURE 1.31   The Response Surface Voltage: 70 V.

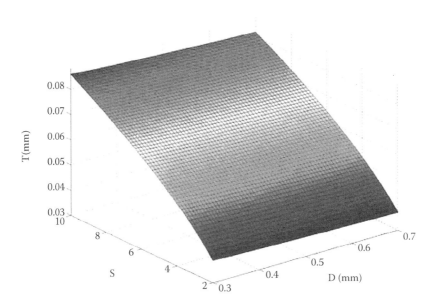

FIGURE 1.32   The Response Surface Voltage: 90 V.

plots can be acquired as following. The thickness change decreases with the distance between electrodes.

2.8.5 CONCLUSION

From the experiment, the following conclusions can be made:

1. The electrical behaviors of current and voltage change were investigated. The current change rate is negative proportional to the experiment time, but the voltage change rate is inversed.
2. The three ELID parameters were applied within the designed range. The theory math model was introduced and predicted that the oxide layer growth rate increased with increasing voltage and velocity of flow, while it decreased with the gap increment. The nonlinear regression model was developed to confirm the theory model.
3. It has been proved that the higher voltage, smaller gap, and higher velocity of flow were helpful in reducing the dressing time during the ELID grinding process.

REFERENCES

1. H. Ohmori and T. Nakagawa, Grinding of Silicon Using Cast Iron Fiber Bonded Wheel (3rd report), *Proceedings of Autumn Conference of JSPE* (1987), 687. (In Japanese)
2. I. Takahashi, H. Ohmori, and T. Nakagawa, Highly Efficient Grinding of Hard and Brittle Materials, *Proceedings of Autumn Conference of JSPE* (1988), 401–402. (In Japanese)
3. H. Ohmori and T. Nakagawa, Mirror Surface Grinding of Silicon Wafers with Electrolytic In-Process Dressing, *CIRP Annals Manufacturing Technology* 39, 1 (1990), 329–332.
4. H. Ohmori, I. Takahashi, and T. Nakagawa, Mirror Surface Grinding by Metal Bonded Superabrasive Wheel with Electrolytic In-Process Dressing, *Progress in Precision Engineering* (1991), 152–158.
5. H. Ohmori, and T. Nakagawa, Electrolytic In-Process Dressing (ELID) for Mirror Surface Grinding, *Proceedings of 10th International Symposium for Electro-Machining* (1992), 553–559.
6. H. Ohmori, K. Katahira, M. Anzai, A. Makinouchi, Y. Yamagata, S. Moriyasu, and W. Lin, Mirror Surface Grinding Characteristics by Ultraprecision Multi-Axis Mirror Surface Machining System, *Journal of JSAT* 45, no. 2 (2001), 85–91. (In Japanese)
7. H. Ohmori, Precision Micro Machining of Ceramics, *Journal of Ceramic Society of Japan* 39, no. 12 (2004), 968–969. (In Japanese)
8. N. Itoh, H. Ohmori, S. Moriyasu, T. Kasai, K. Toshiro, and B. P. Bandyopadhyay, Finishing Characteristics of Brittle Materials by ELID-Lap Grinding Using Metal-Resin Bonded Wheels, *International Journal of Machine Tools and Manufacture* 38 (1998), 747–762.
9. T. J. Vickerstaff, Diamond Dressing: Its Effect on Work Surface Roughness, *Industrial Diamond Review* 30 (1970), 260–267.
10. C. E. Davis, The Dependence of Grinding Wheel Performance on Dressing Procedure, *International Journal of Machine Tool Design Research* 14 (1974), 33–52.
11. W. Konig and H. P. Meyen, AE in Grinding and Dressing: Accuracy and Process Reliability, *SME* (1990), MR 90–526.
12. S. Malkin and T. Murray, Mechanics of Rotary Dressing of Grinding Wheels, *Journal of Engineering for Industry, ASME* 100 (1978), 95–102.

13. P. Koshy, V. K. Jain, and G. K. Lal, A Model for the Topography of Diamond Grinding Wheels, *Wear* 169 (1993), 237–242.
14. K. Syoji, L. Zhou, and S. Mitsui, Studies on Truing and Dressing of Grinding Wheels, 1st Report, *Bulletin of the Japan Society of Precision Engineering* 24, no. 2 (1990), 124–129.
15. K. Suzuki, T. Uematsu, T. Yanase, and T. Nakagawa, On-Machine Electro-Discharge Truing for Metal Bond Diamond Grinding Wheels for Ceramics, *Proceedings of the International Conference on Machining of Advanced Materials, NIST* 847 (1993), 83–88.
16. X. Wang, B. Ying, and W. Liu, EDM Dressing of Fine Grain Super Abrasive Grinding Wheel, *Journal of Materials Processing Technology* 62 (1996), 299–302.
17. M. A. Piscoty, P. J. Davis, T. T. Saito, K. L. Blaedel, and L. Griffith, Use of In-Process EDM Truing to Generate Complex Contours on Metal Bond Superabrasive Grinding Wheels for Precision Grinding Structural Ceramics, *Proceedings of International Conference on Precision Engineering*, Taipei, Taiwan, 1997, 559–564.
18. I. Marinescu, W. B. Rowe, B. Dimitrov, and I. Inasaki, *Tribology of Abrasive Machining Process*, Norwich, NY, William Andrew Inc., 2004.
19. H. Ohmori, I. Takahashi, B. P. Bandyopadyay, Ultra-Precision Grinding of Structural Ceramics by Electrolytic In-Process Dressing Grinding, *Journal of Materials Processing Technology* 57 (1996), 272–277.
20. V. Bagotsky, *Fundamentals of Electrochemistry*, 2nd ed., Hoboken, NJ, Wiley-Interscience, 2005.

2 Fundamental ELID Grinding Types

Hitoshi Ohmori and Kazutoshi Katahira

CONTENTS

2.1 ELID SURFACE GRINDING

Surface grinding, which is the most common type of grinding operation, is the process of fabricating flat-surfaced workpieces using reciprocal or rotary grinding machines equipped with straight-type grinding wheels. Some of these grinding systems use cup wheels and adopt a vertical grinding axis. Figure 2.1 shows a straight-type grinding wheel equipped with electrolytic in-process dressing (ELID) electrodes. In this view, the wheel cover includes the −Ve pole electrode, while the +Ve pole carbon-brush electrode achieves and maintains smooth contact with the wheel center when the cover is shut. The chemical-solution coolant applicator of the ELID function is installed in a way that allows it to supply fluid to both grinding wheels to achieve the ELID function, as well as to cool, clean, and lubricate the grinding points. The electrode can be adjusted manually to achieve a specific gap. Figure 2.2 shows a rotary grinding system with a straight-type grinding wheel. Since the ELID surface grinding system that uses a straight wheel can be applied to the plane grinding requirements of a wide variety of industrial fields, widespread and general-purpose practical use of this technique is expected to increase. Furthermore, as shown in Figure 2.3, the ELID grinding method can also be applied to in-feed grinding systems, in which a workpiece rotates while being ground, and is thus of practical use in a variety of situations.[1-3]

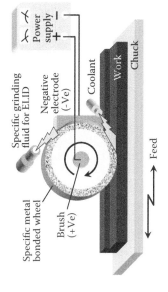

FIGURE 2.1 Surface grinding method.

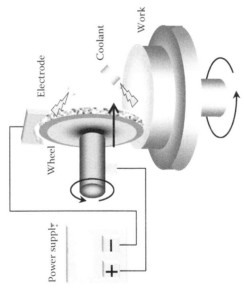

FIGURE 2.2 Rotary grinding system.

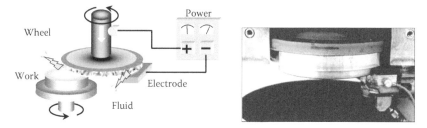

FIGURE 2.3 In-feed grinding systems.

2.2 ELID CYLINDRICAL GRINDING

As described, although research began from the viewpoint of plane grinding using a cup- or straight-type wheel, there is currently a high demand for a high-quality mirror surface finishing technology that can be applied to cylindrical-shaped products. Cylindrical grinding is also called center-type grinding and is used to remove excess material from the cylindrical surfaces and shoulders of workpieces. As shown in Figure 2.4, a cylindrical grinding system is composed of an axis that holds and rotates the cylindrical workpieces, and a wheel axis that grinds the outer circumference of the workpiece. Both the tool and the workpiece are rotated by separate motors and at different speeds. The axes of a rotation tool can be adjusted to produce a variety of shapes. The cylindrical grinding process uses straight-type or cup-type wheels. When used in this process, the ELID system configuration is almost the same as that used in the case of plane grinding. The wheel serves as the +Ve pole by way of a brush electrode that is kept in smooth contact with the rotating center, while the electrode fixed below it becomes the –Ve pole. When supplied with the ELID fluid and an electric current, electrolysis occurs in the small (approximately 0.1 mm) clearance area between the –Ve and the +Ve poles.[4,5]

2.3 ELID INTERNAL GRINDING

The internal grinding method is used to process the interiors of cylindrical workpieces using wheels with diameters that are smaller than the inner diameters of such workpieces. It is difficult to achieve mirror surface internal grinding of such workpieces

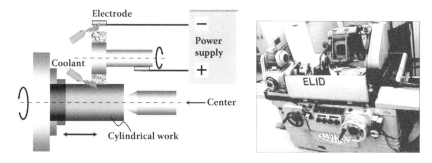

FIGURE 2.4 Application of ELID system to cylindrical grinder.

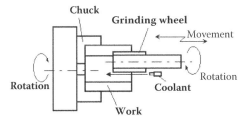

FIGURE 2.5   Application of ELID II system to internal cylindrical grinder.

using conventional grinding technology. The newly developed ELID internal mirror grinding method is a processing system that utilizes a highly streamlined honing method. For internal cylindrical grinding, especially when the diameter of the workpiece is just slightly larger than that of the grinding wheel, it becomes difficult (and sometimes even impossible) to mount a dressing electrode parallel to the wheel as is done in a more common ELID system. To cope with this difficulty, the electrolytic interval dressing technique (ELID II system) shown in Figure 2.5 has been developed. This system uses the abbreviated name of "electrolysis interval dressing" to distinguish it from "in-process" dressing; the term *ELID II* is used here.[6–8]

2.4 ELID CENTERLESS GRINDING

An ELID centerless grinding machine is shown in Figure 2.6. The centerless grinding method grinds the outside diameter of cylindrical workpieces by applying rotary torque from the frictional force of the regulating wheel, without completely immobilizing the small diameter cylindrical workpiece itself. When a wider grinding wheel (50 mm) is used, a higher current and voltage is needed for initial electrolytic dressing. A longer dressing time (approximately 40 minutes) is also required. This method can also be applied to workpieces below 100 μm, and is capable of efficiently grinding a micropin core with a very high aspect ratio.[9]

2.5 ELID CURVE GENERATOR (CG) GRINDING

Similar to the conventional ELID grinding system, ELID curve generator (CG) grinding is essentially composed of the following elements: (a) a cup metal-bonded diamond wheel, (b) an ELID direct current (DC)-pulse power source, (c) a specific grinding fluid (which served as an electrolytic agent), and (d) a fixed copper electrode. Figure 2.7 shows a schematic diagram of the ELID CG grinding process. The generating mechanism of the spherical surface ground with cup wheels was first introduced in 1920 by W. Taylor, an English scholar. As shown in Figure 2.7, the workpiece is mounted on a work spindle, and the inclination angle α between the axis of rotation of the workpiece and that of the wheel spindle is properly adjusted.

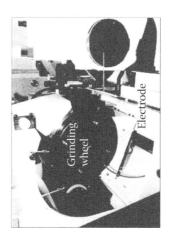

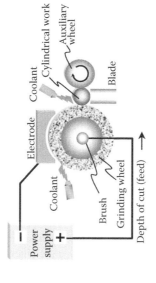

FIGURE 2.6   ELID centerless grinding method.

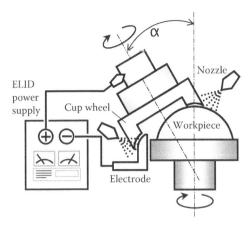

FIGURE 2.7 Schematic of ELID CG grinding.

Theoretically, the radius of curvature of the lenses that are produced can be calculated using the following equations:

$$R = Dsi/(2 \times \sin\alpha) \text{ (for convex lens)} \tag{2.1}$$

$$R = Dso/(2 \times \sin\alpha) \text{ (for concave lens)} \tag{2.2}$$

where R is the radius of curvature of the lenses, Dsi is the internal diameter of the grinding wheel, Dso is the external diameter of the grinding wheel, and α is the inclination angle of the axes of rotation between the workpiece and the wheel.

From a geometrical point of view, the parameters of dimensional error and shape accuracy in ELID CG grinding consist mainly of the following: (a) inclination deviation of the workpiece axis from the wheel axis, (b) position deviation of the wheel–workpiece contact point from the workpiece rotation center in the α plane, and (c) position deviation of the wheel axis from the α plane. Among these, the first two affect the dimensional error of the lenses, and the last one influences the shape accuracy of the lenses. In contrast, wheel wear has no impact on the shape accuracy of the lenses that are produced. Figure 2.8 shows an overview of the experimental setup.

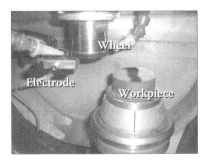

FIGURE 2.8 Overview of experimental setup.

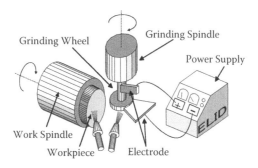

FIGURE 2.9 ELID application for aspheric grinding.

2.6 ELID ASPHERIC GRINDING

The achievement of mirror surface finishing for various hard and brittle materials has been realized with the development of the ELID grinding method. Application of the ELID grinding method is no longer restricted to plane surface grinding; it has expanded to cylindrical grinding, internal grinding, and spherical-shape grinding as well. The demand for this technique in the development of optical elements, including aspheric shaped lens or molds, is currently higher than ever before. Examples of the conventional grinding systems used for optical lens are shown in Figure 2.9. For all such systems, an exclusive processing machine is used and various wheel forms are included. In the case of an ELID grinding system for optical lens production, ELID is applied using the same configuration shown in the processing system displayed in Figure 2.9. Figure 2.10 shows an overview of ELID aspheric grinding.

FIGURE 2.10 Overview of ELID aspheric grinding.

REFERENCES

1. H. Ohmori and T. Nakagawa, Mirror Surface Grinding of Silicon Wafers with Electrolytic In-Process Dressing, *CIRP Annals Manufacturing Technology* 39, no. 1 (1990), 329–333.

2. H. Ohmori and T. Nakagawa, Analysis of Mirror Surface Generation of Hard and Brittle Materials by ELID (Electronic In-Process Dressing) Grinding with Superfine Grain Metallic Bond Wheels, *CIRP Annals Manufacturing Technology* 44 (1995), 287–290.

3. H. Ohmori and T. Nakagawa, Utilization of Nonlinear Conditions in Precision Grinding with ELID (Electrolytic In-Process Dressing) for Fabrication of Hard Material Components, *CIRP Annals Manufacturing Technology* 46 (1997), 261–264.

4. H. Ohmori, S. Moriyasu, W. Li, I. Takahashi, K. Y. Park, N. Itoh, and B. P. Bandyopadhyay, Highly Efficient and Precision Fabrication of Cylindrical Parts from Hard Materials with the Application of ELID (Electrolytic In-Process Dressing), *Materials and Manufacturing Processes* 14 (1999), 1–12.

5. J. Qian, L. Wei, and H. Ohmori, Cylindrical Grinding of Bearing Steel with Electrolytic In-Process Dressing, *Precision Engineering* 24 (2000), 153–159.

6. H. Ohmori and J. Qian, ELID-II Grinding of Micro Spherical Lens, *RIKEN Review* 23 (2000), 140.

7. J. Qian, H. Ohmori, and W. Lin, Internal Mirror Grinding with a Metal/Metal-Resin Bonded Abrasive Wheel, *International Journal of Machine Tools and Manufacture* 41 (2001), 193–208.

8. C. Zhang, H. Ohmori, and W. Li, Small-Hole Machining of Ceramic Material with Electrolytic Interval Dressing (ELID-II) Grinding, *Journal of Materials Processing Technology* 105 (2000), 284–293.

9. H. Ohmori, W. Li, A. Makinouchi, and B. P. Bandyopadhyay, Efficient and Precision Grinding of Small, Hard, and Brittle Cylindrical Parts by the Centerless Grinding Process Combined with Electro-Discharge Truing and Electrolytic In-Process Dressing, *Journal of Materials Processing Technology* 98 (2000), 322–327.

Section II

ELID Operations

3 ELID Grinding Methods

Hitoshi Ohmori and Kazutoshi Katahira

CONTENTS

3.1 ELID I (MIST ELID)

The electrolytic in-process dressing (ELID) grinding system has been successfully introduced into efficiency and high-definition processing technology of hard and brittle material so far. Their areas of use are spreading in the manufacture of electronic and optics components. On the other hand, demands for environmental-friendly grinding technologies are increasing in the manufacturing industry. The authors are therefore studying the possibilities of environmental-friendly ELID grinding technologies from the perspectives of the electrolytic dressing method, grinding wheel, grinding fluid, and so forth.

One way of realizing environmental-friendly processing technology is to minimize the quantity of grinding fluid (MQL, minimum quantity lubrication). Based on this concept, we proposed a semidry ELID grinding method and conducted experiments on the practicality of this method. This chapter reports the results of electrolytic dressing characteristics and grinding characteristics of the semidry ELID grinding method.

Figure 3.1 shows a schematic illustration of the semidry ELID grinding system and Figure 3.2 shows a close-up view. The semidry ELID grinding system is more or less the same as the ELID grinding system, except that the normal grinding fluids

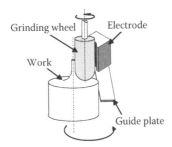

FIGURE 3.1 Schematic illustration of semidry ELID method.

supply method can be changed to mist supply to realize a clean environment. In this grinding system, positive potential is applied to the grinding wheel and negative potential to the electrode. A special pulse generator is used. Mist mixed with a small amount of grinding fluid and compressed air is supplied to the space between the electrode and the grinding stone. Electrolytic dressing is done and ground also.

Figure 3.3 shows the exterior of the machine developed specifically for micro tool machining. The machine is designed to be extremely compact and can be used on a desktop. This machine has three linear axes: X, Y, and Z. In the processing method, the work stage moves in the X and Y directions with respect to the outer circumference of the grinding wheel, which is fixed. This movement accomplishes both feeding and notching of the workpiece, finishing the workpiece to the required shape.[1–5]

A plasma discharge truing system is used before the electrolytic dressing and grinding process in order to achieve the fine shape of grinding wheel. The positive electrode is attached to the grinding wheel and the negative electrode to the truing wheel. The grinding wheel moves along the circumference of the truing wheel. Mist-type grinding fluid is supplied for cooling. The ELID power supply used in this case can be used for ELID grinding as well as plasma discharge truing. Figure 3.4 shows the plasma discharge truing system.

FIGURE 3.2 Close-up view of semidry ELID grinding system.

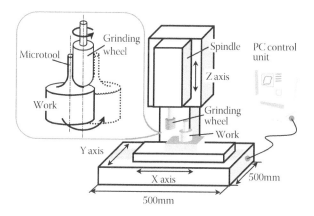

FIGURE 3.3   External view of developed machine for microtool machining.

The experimental system was composed of a grinding machine, electrode, power supply, grinding wheel, and workpiece. The experimental conditions were as follows:

- Grinding machine—Desktop cylindrical grinding machine
- Grinding wheel—#1200 metal-bonded diamond grinding wheel measuring 8 mm in diameter and 10 mm in length.
- ELID power supply—ELID pulse generator (NX-1501)
- Grinding fluid—Electroconductive grinding liquid (CEM) diluted to 50 times with purified water
- Workpiece—Cemented carbide micropin

Other experimental conditions are shown in Table 3.1.

First, electrolytic characteristics were investigated after the grind wheel used was plasma trued and previous electrolytic dressing was completed. Then, the grinding characteristics were studied through a four-pyramid-shaped micropin grinding experiment. The compressed air used to supply mist was 0.2 MPa. The gap between the mist nozzle and electrode was 30 mm and that between the electrode and grinding wheel

FIGURE 3.4   Plasma discharge truing system.

TABLE 3.1
Experimental Conditions

Plasma Truing Conditions

Rotation speed	10000 min^{-1}
Open voltage	Vp: 90 V
Peak current	Ip: 1 A

Electrolytic Dressing Conditions

Rotation speed	8000, 10000, 15000 min^{-1}
Open voltage	Vp: 90 V
Peak current	Ip: 1 A

Grinding Conditions

Rotation speed	8000, 10000, 15000 min^{-1}
Open voltage	Vp: 90 V
Peak current	Ip: 1 A
Cutting depth	0.5 µm

was 70 µm. The mist supplied was minimized within 6.4 and 49.0 ml/min to realize the electrolytic dressing. Incidentally, the normal quantity grinding fluid supplied for ELID grinding is around 1000 ml/min. The quantity of grinding fluid supplied was successfully reduced by more than 90%.

The mist was supplied within three conditions during electrolytic dressing: 49.0 ml/min, 27 ml/min, and 6.4 ml/min. Figure 3.5 shows the surface condition of the grinding wheel before electrolytic dressing. The surface of the grinding wheel became yellowish after electrolytic dressing. It suggests that the electrolytic film was formed during electrolytic dressing.

Figure 3.6 shows the relation between the thickness of electrolytic film and quantity of grinding fluid when the grinding wheel was rotated at a speed of 15000 min^{-1}

FIGURE 3.5   After electrolytic dressing.

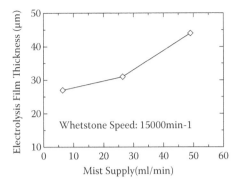

FIGURE 3.6   Relation between electrolysis film thickness and mist supply.

for 30 minutes. The tendency that the electrolytic film becomes thicker with increasing of grinding fluid quantity is also shown. It can be considered that grinding fluid between the electrode and the whetstone easily gather and stable electrolytic dressing becomes easier when grinding fluid quantity is increased. Even other rotation speeds showed a similar tendency.

Figure 3.7 shows the relation between the thickness of electrolytic film and rotation speed of grinding wheel when mist was supplied at 6.4 ml/min for 30 minutes. The tendency that the thickness of electrolytic film becomes thinner with increasing of grinding wheel rotation speed can be seen. The influence of the centrifugal force increases and the grinding fluid adhered to the grinding wheel surface can be more easily removed at a higher grinding wheel rotation speed. This suggests that the mist supply required for the electrolytic dressing is insufficient. Also at the grinding wheel rotation speed 20,000 min^{-1}, electric discharge is generated, disabling stable electrolytic dressing. This may be because the grinding fluid was insufficient, and further studies on changing the shape of the electrode, nozzle distance, and mist supply method are required.

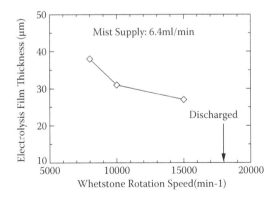

FIGURE 3.7   Relation between electrolysis film thickness and whetstone rotation speed.

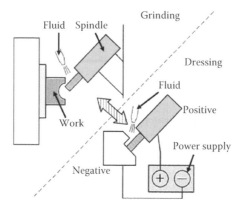

FIGURE 3.8 ELID II method.

3.2 ELID II (INTERVAL ELID)

The ELID II method is electrolytic interval dressing. This means that the electrolytic dressing is not applied with in-process but with interval. This method is good for application of small grinding wheels that have difficulties employing a small electrode with a small clearance. So, in these cases, grinding without electrolytic in-process dressing and electrolytic dressing are repeated with some interval. Figure 3.8 shows the schematic of this principle. In recent years, demands on microlens and their molds are being rapidly increased. Figure 3.9 shows the ELID II system for

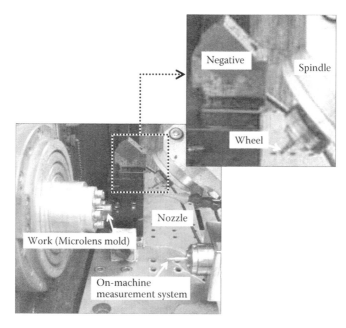

FIGURE 3.9 ELID II system for microlens mold fabrication.

microlens mold fabrication. A very small grinding wheel is tilted and approached to concave work grinding, and before and after the grinding, electrolytic dressing is applied with some interval.

The ELID II method is also applied for the honing process. This process needs the entire wheel surface to be in contact with the workpiece's internal surface. So, electrolytic in-process dressing is impossible. Electric discharge truing, electrolytic predressing, and then the ELID II honing are repeatedly conducted.

3.3 ELID III (ELECTRODELESS ELID)

Conventional ELID grinding needs space for setting up the electrode to dress the grinding wheels outside the workpieces. To apply ELID grinding to the machining of certain types of grinding wheels with special shapes, interval dressing (ELID II) had to performed. The purpose of the development of the ELID III grinding method was to construct a new electrodeless electrolytic dressing system.

The ELID III grinding method is an electrodeless electrolytic dressing system. As shown in Figure 3.10, the negative pole of the power supply is connected to the workpiece that is limited to conductive materials. The conductive wheel serves as the positive pole, while the conductive workpiece serves as the negative pole. The conductive wheels are made of semiconductors composed of metal and plastic powder. The workpieces are made of such metallic materials as steels. This combination of semiconductor wheels and metallic workpieces is ideal for preventing electrodischarge and for producing mild electrolytic dressing effects between these poles. Figure 3.11 shows the principle of the ELID III grinding method.

The basic grinding characteristics of ELID III grinding by a lapping machine was investigated, followed by the machining characteristics of a curved surface

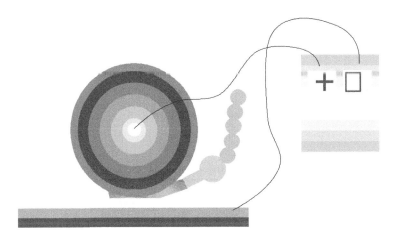

FIGURE 3.10 ELID III method.

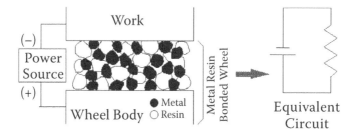

FIGURE 3.11   Principle of ELID III grinding method.

using ball-nose wheels and changes in the current during grinding and surface conditions of the wheel. After machining, the surface roughness of each workpiece was measured.

Table 3.2 shows the experimental system. Table 3.3 shows the grinding conditions by the lapping machine. Table 3.4 shows the grinding conditions using ball-nose wheels. Figure 3.12 shows the view of the experimental grinding system. Figure 3.13

TABLE 3.2
Experimental System

Grinding machine	Single Sided Lapping Machine: MG-773B	NC Milling Machine: FNC-105
Grinding wheel	Metal-resin bonded (Metal:resin = 7:3) SD1200	Metal-resin bonded (Metal:resin = 7:3) Size: D20 × R10 × T20 × L120 SD80, SD200
Power source	NX-ED630	SPS-01
Grinding fluid	NX-CL-CM2	
Surface measurement	Surftest-701	

TABLE 3.3
Grinding Conditions by Lapping Machine

	Silicon	S35C
Wheel rotation speed (rpm)	110	110
Work rotation speed (rpm)	80	80
Pressure (kgf)	10.6–16.0	10.6
Open voltage (V)	60	20
Peak current (A)	30	4
On/off time (micro second)	2	2

TABLE 3.4

Grinding Conditions Using Ball-Nose Wheels

	#80	#200
Rotation speed (rpm)	1000	1000
Feeding (X) (mm/min)	200–400	200
Depth of cut (Y) (mm)	0.010–0.020	0.005–0.015
Open voltage (V)	20–60	20
Peak current (A)	10	10
On/off time (micro s)	2	2

FIGURE 3.12 View of lapping machine.

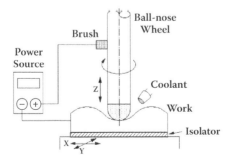

FIGURE 3.13 ELID III grinding method using ball-nose wheels.

TABLE 3.5
Surface Roughness

	Ra (nm)	Rz (nm)	Ry (nm)
Silicon	49.8	300.2	426.0
S35C	23.6	104.0	167.0

shows the ELID III grinding method using a ball-nose wheel. The workpieces used were silicon (82 mm across), S35C (50 mm across), and SKD11 (50 × 20 × 50 mm).

1. ELID III grinding—In the case of silicon, changes to the current were confirmed, and an oxide layer was formed on the surface wheel. Impurities were formed on the surface workpiece when the pressure was set to 10.6 kgf. Better surface roughness was obtained when the pressure was set to 16 kgf. In the case of S35C, an oxide layer was formed on the wheel surface when the peak current was 4 A and the open voltage was 20 V. No oxide layer was formed for silicon at these values. The actual current was formed to improve about 1.0 to 1.5 A for both cases.

2. Surface roughness—The surface roughness obtained by ELID lap grinding for silicon using the #1200 wheel was about 600 nm Ry. Better surface roughness was achieved by ELID III grinding, and mirror surface finish was also realized for S35C. Table 3.5 shows the surface roughness after grinding. Figure 3.14 shows the surface samples of silicon and S35C finished by ELID III grinding.

3. Ball-nose wheels—A spark phenomenon occurred between the wheel and workpiece when the peak current was 10 A and the open voltage was 60 V, which results in the damage of the wheel and workpiece. An oxide layer was formed on the wheel surface when the peak current was 10 A and the open voltage was 20 V. Better grinding was achieved using ball-nose wheels. Better machining was achieved when the depth of cut was under 0.010 mm using the #80 wheel and when the depth of cut was under 0.005 mm using the #200 wheel. Figure 3.15 shows the surface of the ball-nose wheels.

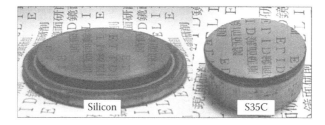

FIGURE 3.14   Surface samples of silicon and S35C finished by ELID III grinding.

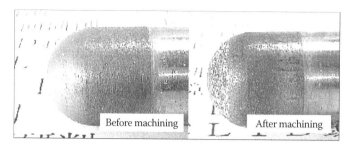

FIGURE 3.15   Surface of the ball-nose wheels.

3.4   ELID IV (ION-SHOT ELID)

The ELID IV method is the ion-shot in-process dressing. The grinding fluid is dissolved between the positive and negative poles on the top area of the grinding fluid nozzle.[6–9] This method is also called the nozzle ELID method. This method does not use any electrode around the grinding wheel surface, so the entire grinding wheel surface is free for grinding operation. Figures 3.16, 3.17, and 3.18 show its principle, a desktop machine for ELID IV, and machined lens molds, respectively.

In the fundamental experiments, the ultraprecision desktop 4-axis control grinding machine TRIDER-X was used. The workpiece used was CVD-SiC, and the

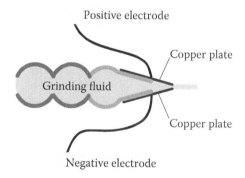

FIGURE 3.16   ELID IV method.

FIGURE 3.17   Desktop ELID IV grinding system.

(a) State of grinding

(b) Ground lens mold (CVD-SiC)

FIGURE 3.18   Grinding scene in (a) and (b).

grinding tool was a cast-iron bond grinding wheel (#4000). The experiment aimed through the application of the ion-shot dressing grinding system to process an aspheric lens mold. The workpiece used CVD-SiC coating for the matrix of SiC. Experimental conditions are listed in Table 3.6. Surface roughness is represented by

TABLE 3.6
Experimental Conditions

Grinding wheel	Bond material: Cast-iron bond (FCI-X)
	SD#4000
	Shape: φ75 mm, 5 mm in thickness
Tool rotational speed (min⁻¹)	2500
Depth of cut (mm/pass)	0.001
Total depth of cut (mm)	0.1
Feed rate (mm/min)	1
Work rotational speed (min⁻¹)	10
Grinding fluid	NX-CL-CG7
	(50 times dilution)
Electrolytic conditions	Vp(V): 150
	Ip(A): 60
	On time (μs): 2
	Off time (μs): 2

Ra in Figure 3.19a. Ra of 1.5 nm and good surface roughness were produced. In addition, form accuracy was achieved for the first processing of 3.4 μm in Figure 3.19b. At the second processing, when the center position of the grinding wheel was corrected, the result was P-V 0.79 μm. This realizes form accuracy of less than 1 μm. Thus, the ion-shot dressing grinding system achieved a dressing effect superior to that produced by an ELID system.

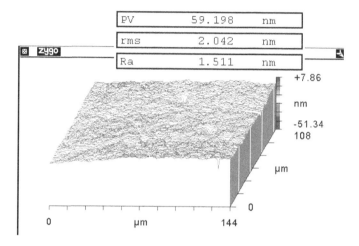

PV	59.198	nm
rms	2.042	nm
Ra	1.511	nm

(a) Surface roughness (ZYGO NewView optical profiler)

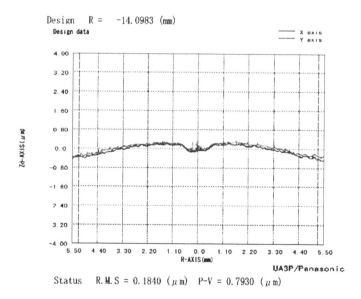

(b) Form accuracy (Panasonic UA3P, Ultrahigh Accurate three-dimensional profilometer)

FIGURE 3.19 Result for the lens glass mold. (a) Surface roughness (new view). (b) Form accuracy (Panasonic UA3P, Ultrahigh Accurate three-dimensional profilometer).

3.5 ACOUSTIC EMISSION (AE) MONITORING OF ELID GRINDING

3.5.1 INTRODUCTION

Sapphire, a single-crystal form of α-alumina, is widely used in a range of high-technology applications, such as optics, electronics, and high temperature sensors, because of its combination of excellent optical, mechanical, physical, and chemical properties.[10–13]

In many of these applications, super surface finish and high accuracy are required. Currently, the machining of sapphire is mainly done by fine grinding, lapping, or polishing. Compared with lapping and polishing, fine grinding can be used to efficiently machine sapphire at a relatively low cost. Metal-bonded superabrasive grinding wheels are widely used to obtain a mirror surface finish on brittle and hard materials. However, fine grinding of sapphire is quite challenging because of its high hardness and low fracture toughness, making it sensitive to cracking. In addition, low grinding efficiency due to wheel loading and rapid abrasive wear that requires periodic wheel dressing is another major problem in conventional grinding of sapphire. To reduce the surface roughness and subsurface damage, superabrasive grinding wheels with smaller diamond grains are needed. However, such wheels are susceptible to loading and glazing, especially if fine grains are used.[14–16] Periodic dressing is required to minimize the aforementioned problems, which makes the grinding process less efficient.

ELID grinding shows great promise in overcoming the problems of conventional grinding of hard and brittle materials. This technology provides dressing of the metal-bonded wheels during the grinding process, while maintaining sharp abrasives from the superabrasive wheels.[17–19] Over the past 20 years, ELID has been used to grind various hard and brittle materials. It is found that this method is highly efficient in grinding hard and brittle materials. It has been reported that ductile-mode grinding with smoother surface finish and lower subsurface damage was obtained on silicon wafers by using superabrasive and the ELID technique.[20–21] The improvement of surface finish was explained by the formation of a beneficial oxide layer on the grinding wheel surface. The oxide layer acts as a spring damper, which reduces the grit depth of cut and absorbs the vibrations in the ELID grinding process.[14,16] Force was commonly used by previous research to monitor the ELID grinding process and to investigate its mechanisms. A significant reduction of grinding force has been reported with the application of ELID.[17] Lim et al. investigated the influence of dressing parameters on ELID grinding and pointed out that the grinding force is reduced with the increase of dressing current duty ratio. Furthermore, he also found that the tangential force is unstable due to the macrofracture of the oxide layer from the wheel surface.[16]

Acoustic emission (AE) monitoring of the grinding process was investigated by Dornfeld and Cai.[22] They discovered that the AE RMS (root mean square) is most strongly influenced by the chip thickness determined by the depth of cut, and that it is more sensitive than force measurement to detect the wheel–workpiece contact and spark-out. They also showed that the AE energy increases with the increase of wheel loading and wear. This was further confirmed by Inasaki and Okamura.[23] Stephenson et al. studied the AE RMS and AE raw signals in monitoring ELID grinding of BK7 glass and Zerodur.[25] Higher AE energy was found at gentle dressing parameters in ELID grinding. The main advantage of AE monitoring during the grinding process is that the frequency of AE signals ranges from 20 KHz to 1 MHz, which is above most structural natural frequencies, such as frequencies of machine vibrations and other mechanical noises. It is known that the relatively reliable parameter in AE monitoring of the grinding process is the AE RMS. However, the main limitation in AE monitoring by RMS is unpredictable change in signal linearity and amplitude with time.[24] The fluctuation of AE RMS may be caused by minor and unpredictable changes in grinding conditions. To overcome this problem, an innovative grit mapping technique was proposed by Gomes de Oliveira and Dornfeld to give information about the grinding wheel topography and the dressing operation by using fast AE RMS analysis to characterize and map each abrasive-tool contact.[24]

In this study, the AE technique was used to monitor ELID grinding of sapphire. The influence of grinding conditions and dressing parameters on surface finish and acoustic emission signals were investigated. AE RMS was used as the parameter to analyze the AE signals generated in the ELID grinding process of sapphire.

3.5.2 EXPERIMENTAL SETUP

The experiment was performed on a Thompson Creep Feed Grinder equipped with Fanuc CNC, an ELID system, a dynamic balance system, and an acoustic emission system. A copper electrode, covering one-sixth of the perimeter of the grinding wheel, was used for electrolytic dressing. A cast-iron bond diamond grinding wheel was connected to the positive pole through the contact of a carbon brush. The specification of the wheel was diameter 305 mm, width 11 mm, and grit size #4000 (average grain size 4 µm). The gap between the wheel and electrode was maintained at 0.2 mm. A special coolant, TRIM C270, was diluted with distilled water to a ratio of 1:20 and used as an electrolyte and coolant for the experiment. A direct current pulse generator was used as the power supply to generate a square pulse wave for the ELID process. The surface finish was measured by the HOMMEL-TESTER T 1000. The method of measuring the surface roughness is the stylus method. The AE testing was performed with the aid of the MISTRAS 2001 system, which was developed by the Physical Acoustic Corporation. A piezoelectric type transducer, R15, was used in the study. The sensor was attached to the surface of the workpiece holder. The acoustic emission signals detected by the sensor were amplified to usable voltage levels by a 2/4/6 type preamplifier. The 2/4/6 type preamplifier was set at a gain of 100 (40 dB). The detected AE signals by the sensor were first amplified by the preamplifier and then fed to the AEDSP-32/16 board where the processing of the raw AE signals was done and stored for later

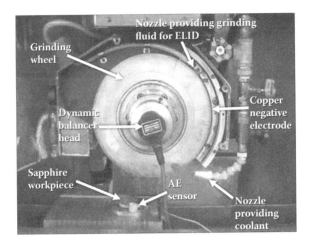

FIGURE 3.20 Configuration of experimental setup.

analysis. A frequency of 4 MHz sample rate was selected for signal acquisition. The experimental configuration is shown in Figure 3.20.

3.5.3 RESULTS AND DISCUSSION

A series of ELID grinding tests were conducted to investigate the effect of grinding conditions on surface finish of the ground sapphire workpiece and AE signals under the following conditions:

- Wheel speed—16 m/s (1000 rpm)
- Feed rate—0.5 m/min
- Depth of cut—1 μm, 3 μm, 5 μm, 7 μm, 9 μm, 11 μm, 13 μm, 15 μm
- Initial ELID current—5 A, 7 A, 9 A

3.5.3.1 Predressing and Oxide Layer Thickness

ELID predressing of the grinding wheel is a preliminary stage, which is required before starting ELID grinding. During the ELID predressing stage, the abrasive grains are protruded by removing the metal bond material from the wheel surface through electrolysis. At the same time, a beneficial oxide layer forms on the wheel surface. The diameter of the grinding wheel increases after predressing due to the formation of this oxide layer. In the case of a cast-iron bonded grinding wheel, the cast-iron bond is first ionized into Fe^{2+}. The ionized Fe then reacts with OH^- to form hydroxides, which further transform into oxide (Fe_2O_3), forming an insulating oxide layer on the grinding wheel surface. This insulating layer causes a decrease in current flow and an increase in voltage known as the anode effect. The oxide layer growth rate decreases with the decrease in the current flow, and the electrical resistance in the electrolytic circuit increases with the increase of oxide layer thickness. The oxide layer formed on the cast-iron bonded wheel surface is a yellow-brown layer, as shown in Figure 3.21.

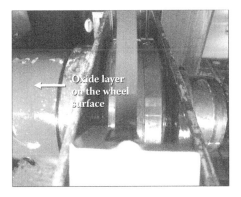

FIGURE 3.21 Yellow-brown oxide layer formed on a cast-iron bonded wheel surface.

The oxide layer thickness during ELID predressing was investigated at three predressing currents. The initial dressing current was 5 A, 7 A, and 9 A, respectively. The gap between the grinding wheel and the negative electrode was maintained at 0.2 mm. The applied voltage was fixed at 60 V. The cast-iron bonded grinding wheel was predressed for 15 minutes in all experiments. The oxide layer thickness is defined as the value of the wheel radius with the oxide layer minus that without the oxide layer.[26] The oxide layer thickness was measured by reading the X coordinates of the CNC grinder before and after a period of electrolytic dressing when the grinding wheel surface touched the same workpiece surface. The AE system was used to detect the contact between the wheel surface and the workpiece surface. The results show that the oxide layer thickness generated in ELID predressing increases with the increase of dressing current, as shown in Figure 3.22. According to Faraday's law, the dissolution rate of a metal anode in electrolysis is proportional to the applied current. Clearly, the current is the determining parameter in the metal removal and

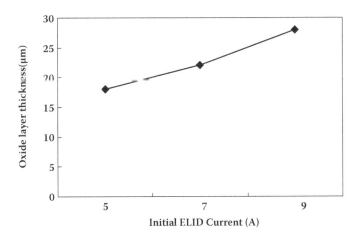

FIGURE 3.22 Effect of electrolytic dressing current on oxide layer thickness.

oxide layer formation processes. After dressing, the cast-iron bond on the wheel surface is replaced by its softer oxide (Fe₂O₃).

During grinding, the insulating layer and abrasive grains are scraped off and removed, resulting in a decrease of the wheel's electrical resistance. As a result, the electrical current through the wheel and the electrode increases and the electrolysis restarts. This dressing cycle repeats during the ELID grinding process. In addition to continuously exposing new sharp diamond grains, ELID can also improve the surface finish quality of the ground workpiece.

3.5.3.2 Effect of Electrolytic Dressing Current on Surface Finish

A series of ELID grinding trials were conducted under the aforementioned grinding and predressing conditions. The average surface roughness (Ra) was measured for each grinding test. The influence of initial ELID currents on the average surface roughness of the ground sapphire workpiece is shown in Figure 3.23. As shown in this figure, increasing the ELID current slightly improves the surface finish quality of the ground sapphire workpiece. Because of the soft oxide layer that formed on the wheel surface, the active grits are bonded in metal oxide matrix that has a lower strength than the original metal bond. The soft oxide layer acts as a spring damper, which plays a role in absorbing the vibration and reducing the grit depth of cut during ELID grinding. The chip thickness model in ELID grinding can be modified by adding a constant (k) to the existing model. h_m can be expressed as[14]

$$h_m = k \left[\frac{4}{Cr} \frac{v_w}{v_s} \left(\frac{a_e}{d_e} \right)^{1/2} \right]^{1/2} \tag{3.1}$$

where k is the ELID constant that ranges from 0 to 1, and a function of dressing current, voltage, current duty ratio, and the coolant properties. The experimental

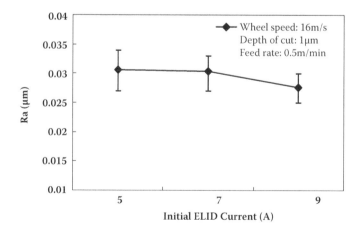

FIGURE 3.23 Effect of electrolytic dressing current on the average surface roughness.

results show that increasing the electrolytic dressing current increases the formation rate and thickness of the oxide layer on the wheel surface, which consequently results in a lower value of k and chip thickness. ELID grinding with thicker oxide layer produces smoother surface finish. It is known that ductile-mode grinding can be achieved when the maximum chip thickness or grit depth of cut is less than the critical depth of cut. The calculated critical depth of cut for sapphire is approximately 30 nm. The application of ELID in grinding of sapphire provides a method to control the grit depth of cut. By controlling the dressing parameters, it is possible to achieve a ductile-mode grinding by using a superabrasive grinding wheel and the ELID technique, which could eliminate the requirement of finishing processes such as lapping and polishing.

3.5.3.3   Effect of Electrolytic Dressing Current on AE Signals

From the aforementioned experimental results, we know the influence of dressing current on the oxide layer thickness and its role on surface finish. This section investigates the effect of dressing current on AE signals. The ELID grinding trials were monitored by the AE system. Figure 3.24 plots the influence of initial ELID currents on the average AE RMS. The average AE RMS was slightly affected by the dressing current. The average AE RMS value was found to be lower at higher dressing current. These results indicate that AE energy decreases with an increase in dressing current. This is because the oxide layer formed on the wheel surface reduces the bond strength and the grit depth of cut during ELID grinding. The increase in dressing current increases the oxide layer thickness and decreases the chip thickness. Since the AE RMS is a function of chip thickness, the increase in ELID current decreases the average AE RMS.

3.5.3.4   Effect of Depth of Cut on Surface Finish and AE Signals

This section investigates the influence of depth of cut on surface finish of the ground sapphire workpiece and the AE RMS during ELID grinding. Figure 3.25a plots the

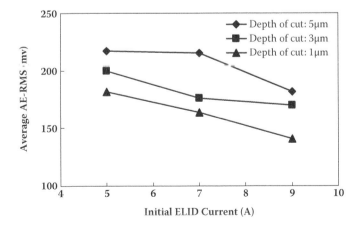

FIGURE 3.24   Effect of electrolytic dressing current on average AE RMS.

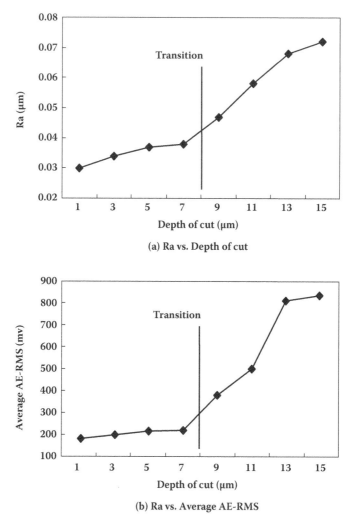

(a) Ra vs. Depth of cut

(b) Ra vs. Average AE-RMS

FIGURE 3.25   Influence of depth of cut on surface finish and average AE RMS. (a) Ra versus depth of cut. (b) Ra versus average AE RMS.

influence of depth of cut on surface finish of sapphire at a wheel speed of 16 m/s, a feed rate of 0.5 m/min, and an initial ELID current of 5 A over a range of depth of cuts (a = 1 μm to 15 μm). The average AE RMS during grinding is shown in Figure 3.25b. The surface roughness and average AE RMS generated during ELID grinding was found to be strongly affected by depth of cut. They both increase with an increase in depth of cut because they are proportional to the chip thickness determined by depth of cut. Furthermore, both figures show that there is a transition point between a = 7 μm and a = 9 μm. Increasing the depth of cut from 1 μm to 7 μm slightly increases the surface roughness of the ground sapphire workpiece. However, the value of Ra increases more quickly with the increase of grinding depth of cut from a = 7 μm to a = 15 μm. The AE RMS shows the same trend.

It has been identified by Lim et al. that the dominant wear mechanism, when the bond strength is reduced during ELID grinding, is bond fracture, which is characterized as grain pullout and oxide layer breakage from the wheel surface.[16] The transition point may be caused by the combined effects of the oxide layer breakage and the subsequent transition of grinding mechanisms from ductile-mode grinding to brittle-mode grinding due to the high depth of cut and oxide layer breakage. Ductile-mode grinding can be realized in ELID grinding under two conditions: a small grinding depth of cut and a relatively thick oxide layer. The smoother surface finish obtained at low depth of cut may be due to ductile-mode grinding. It can be seen from Figure 3.25 that the surface roughness and AE RMS increase more quickly at grinding depths of cut higher than 7 μm because the oxide layer breakage at high depth of cut occurs, resulting in a transition of the grinding mechanisms from ductile-mode grinding to brittle-mode grinding.

The oxide layer breakage can be further verified by examining the electrical current in the ELID grinding process. After the ELID grinding starts, the oxide layer and the abrasives bonded in the metal oxide matrix begin to wear due to the interactions between the oxide layer and the sapphire workpiece, which consequently results in an increase in the electrical current through the wheel and the electrode. The wear rate of the oxide layer depends on the grinding conditions. When the grinding depth of cut is low, the wear rate of the oxide layer and abrasive grains are small due to the gentle grinding force. The breakage of the oxide layer will take place at a relatively high depth of cut due to the large grinding force. As shown in Figure 3.26, the current flow increases slowly when the grinding depth of cut is less than 7 μm. However, the electrical current increases rapidly at a grinding depth of cut higher than 7 μm due to the breakage of the oxide layer from the wheel surface.

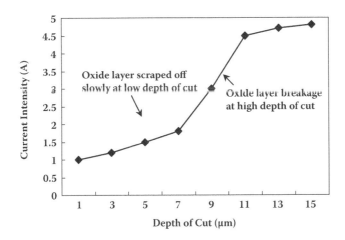

FIGURE 3.26 Dressing current change in ELID grinding with the increase of depth of cut.

A key issue in ELID grinding is to sustain the balance between the removal rate of the bond metal by electrolysis and the wear rate of diamond abrasives.[27] In the grinding process, there are two major forces acting opposite each other on the grit: the grit holding force exerted by the bond material of the grinding wheel and the grinding force. The grinding force increases with an increase in depth of cut. When the grinding force is larger than the grit holding force, the grits will be pulled out. Compared to the abrasives in conventional grinding, the abrasives in ELID grinding are more susceptible to pullout due to the reduced bond strength by electrolysis. Due to the formation of an oxide layer in ELID grinding, the active grits on the wheel surface are bonded in metal oxide matrix that has a lower strength than the original metal bond. ELID grinding at a large depth of cut will increase the chance of grain pullout and oxide layer breakage from the wheel surface. This phenomenon can be detected by AE signals characterized as a rapid increase in the AE RMS value. The oxide layer breakage can also be noticed by examining the electrical current in ELID grinding. A rapid increase in the electrical current indicates the oxide layer breakage. The oxide layer breakage indicates that the wear rate of the oxide layer and abrasives is much higher than the electrolytic dressing rate. Therefore, in order to sustain a relatively stable ELID grinding process, it is not proper to perform ELID grinding at a depth of cut that could cause the oxide layer breakage.

3.5.4   Conclusions

Based on the experimental results, several conclusions can be obtained from ELID grinding of sapphire:

1. In ELID grinding, the soft oxide layer acts as a spring damper, which plays a role in absorbing the vibration and reducing the grit depth of cut. Increasing dressing current increases the oxide layer formed on the wheel surface. The surface finish also improves as the dressing current increases.
2. The AE RMS was found to be affected by the electrolytic dressing current. The AE RMS value was observed to be higher at lower ELID currents.
3. The average AE RMS increases with an increase in depth of cut because the average AE RMS is proportional to the chip thickness determined by depth of cut.
4. The oxide layer breakage occurs at high depths of cut in ELID grinding. The breakage causes rapid increase of the AE RMS value, which indicates the transition of the grinding mechanisms.

REFERENCES

1. H. Ohmori, K. Katahira, Y. Uehara, Y. Watanabe, and W. Lin, Improvement of Mechanical Strength of Micro Tools by Controlling Surface Characteristics, *CIRP Annals Manufacturing Technology* 52 (2003), 467–470.
2. H. Ohmori, K. Katahira, Y. Uehara, and W. Lin, ELID Grinding of Microtool and Applications to Fabrication of Microcomponents, *International Journal of Materials and Product Technology* 18 (2003), 498–508.

3. H. Ohmori, S. Yin, W. Lin, Y. Uehara, S. Morita, M. Asami, and M. Ohmori, Development on Micro Precision Truing Method of ELID-Grinding Wheel (1st Report: Principle and Fundamental Experiments), *Key Engineering Materials* 291/292 (2005), 207–212.

4. Y. Uehara, H. Ohmori, W. Lin, K. Katahira, T. Suzuki, and N. Mitsuishi, Desk-Top Fabrication System, *International Progress on Advanced Optics and Sensors: Proceedings of International Workshop on Extreme Optics and Sensors* (FSS-40), Tokyo, 2003-1 (2003), 241–244.

5. H. Ohmori, Y. Uehara, K. Katahira, Y. Watanabe, T. Suzuki, W. Lin, and N. Mitsuishi, Advanced Desktop Manufacturing System for Micro-Mechanical Fabrication, *Proceedings of Laser Metrology and Machine Performance* VII (2005), 16–29.

6. Y. Uehara, H. Ohmori, W. Lin, Y. Ueno, T. Naruse, N. Mitsuishi, S. Ishikawa, and T. Miura, *Development of Spherical Lens ELID Grinding System by Desk-Top 4-Axes Machine Tool*, 3rd International Conference on Leading Edge Manufacturing in 21st Century (LEM21), Nagoya, Japan, October 2005.

7. Y. Uehara, H. Ohmori, K. Katahira, W. Lin, Y. Watanabe, M. Asami, and N. Mitsuishi, *Advanced Desktop Manufacturing System for Micro-Mechanical Fabrication*, International Conference on Precision Engineering and Micro/Nano Technology in Asia (ASPEN 2005), Shenzhen, China, November 2005.

8. H. Ohmori, Y. Uehara, W. Lin, Y. Watanabe, K. Katahira, T. Naruse, and N. Mitsuishi, *Development of a New ELID Grinding Method for Desk-Top Fabrication System*, International Conference on Precision Engineering and Micro/Nano Technology in Asia (ASPEN 2005), Shenzhen, China, November 2005.

9. Y. Uehara, H. Ohmori, W. Lin, Y. Watanabe, K. Katahira, T. Naruse, N. Mitsuishi, and T. Miura, *Ion Shot Dressing Grinding for Desk-Top Machine Tools with V-Cam System*, 4th International Conference on Leading Edge Manufacturing in 21st Century (LEM21), Fukuoka, Japan, November 2007.

10. H. Zhu and A. Luiz (2004). Chemical Mechanical Polishing (CMP) Anisotropy in Sapphire, *Applied Surface Science* 236 (2004), 120–130.

11. P. K. Chandra and F. Schmid, Growth of the World's Largest Sapphire Crystals, *Journal of Crystal Growth* 225 (2001), 572–579.

12. Y. Wang and S. Liu, Effects of Surface Treatment on Sapphire Substrates, *Journal of Crystal Growth* 274 (2005), 241–245.

13. E. Dorre and H. Hubner, *Alumina: Processing, Properties, and Applications*, New York, Springer, 1984.

14. K. Fathima, K. A. Senthil, M. Rahman, and H. S. Lim, A Study on Wear Mechanism and Wear Reduction Strategies in Grinding Wheels Used for ELID Grinding, *Wear* 254 (2003), 1247–1255.

15. J. H. Liu and Z. J. Pei, ELID Grinding of Silicon Wafers: A Literature Review, *International Journal of Machine Tools and Manufacture* 47 (2007), 529–536.

16. H. S. Lim, K. Fathima, K. A. Senthil, and M. Rahman, A Fundamental Study on the Mechanism of Electrolytic In-Process Dressing (ELID) Grinding, *International Journal of Machine Tools and Manufacture* 42 (2002), 935–943.

17. B. P. Bandyopadhyay and H. Ohmori, Efficient and Stable Grinding of Ceramics by Electrolytic In-Process Dressing (ELID), *Journal of Materials Processing Technology* 66 (1997), 18–24.

18. I. D. Marinescu, *Handbook of Advanced Ceramics Machining*, Boca Raton, FL, CRC Press, 2007.

19. H. Ohmori, I. Takahashi, and B. P. Bandyopadhyay, Ultra-Precision Grinding of Structural Ceramics by Electrolytic In-Process Dressing (ELID) Grinding, *Journal of Materials Processing Technology* 57 (1996), 272–277.

20. B. P. Bandyopadhyay and H. Ohmori, The Effect of ELID Grinding on the Flexure Strength of Silicon Nitride, *International Journal of Machine Tools and Manufacture* 39 (1999), 839–853.
21. J. Q. Gu et al., A Study on the Electrolytic In-Process Dressing (ELID) for Silica Glass in a Ductile Mode, *Journal of Non-Crystalline Solids* 354 (2008), 1398–1340.
22. D. Dornfeld and H. Cai, An Investigation of Grinding and Wheel Loading Using Acoustic Emission, *Transactions of ASME* 106 (1984), 28–33.
23. I. Inasaki and K. Okamura, Monitoring of Dressing and Grinding Process with Acoustic Emission, *CIRP Annals Manufacturing Technology* 34 (1989), 277–280.
24. J. F. Gomes de Oliveira and D. A. Dornfeld, Application of AE Contact Sensing in Reliable Grinding Monitoring, *CIRP Annals Manufacturing Technology* 50, no. 1 (2001), 217–220.
25. D. J. Stephenson, X. Sun, and C. Zervos, A Study on ElID Ultraprecision Grinding of Optical Glass with Acoustic Emission, *International Journal of Machine Tools and Manufacture* 46 (2006), 1053–1063.
26. C. Zhang, H, Ohmori, and W. Li, Small-Hole Machining of Ceramic Material with Electolytic Interval-Dressing (ELID-II) Grinding, *Journal of Materials Processing Technology* 105 (1999), 284–293.
27. H. Chen and J. C. M. Li, Anodic Metal Matrix Removal Rate in Electrolytic In-Process Dressing, *Journal of Applied Physics* 87, no. 6 (2000), 3151–3158.

4 ELID Lap Grinding

Nobuhide Itoh and Hitoshi Ohmori

CONTENTS

4.1 ELID LAP GRINDING (SINGLE SIDED)

4.1.1 INTRODUCTION

Electrolytic in-process dressing (ELID) single-side lap grinding,[1–8] a machining process that employs constant pressure and uses a metal-bonded abrasive wheel with ELID, produces better results in both surface roughness and flatness than other grinding systems. Highly useful for grinding materials for which mirror surface finish cannot be obtained by constant feed grinding, this technique can also apply grain metal-bonded wheels finer than the #1000 wheel. It also has the important advantage of being able to be utilized immediately by mounting onto an existing lapping machine.

4.1.2 ELID SINGLE-SIDE LAP GRINDING SYSTEM

This section describes the effects of ELID single-side lap grinding. Monocrystalline silicon and tungsten carbide were finished by ELID single-side lap grinding using #1200 to #8000 cast-iron bonded diamond wheels (#1200 CIB-D to #8000 CIB-D

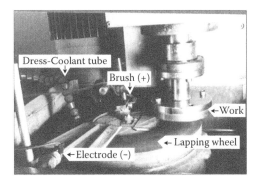

FIGURE 4.1   Close-up view of ELID single-side lap grinding.

wheel), and the effects of mesh number on removal rate and surface qualities of workpieces are described. A simultaneous grinding method of different materials was also introduced.

Figure 4.1 shows a close-up view of an ELID single-side lap grinding machine. In this type of grinding process, the work is pressed onto the lapping wheel. The lapping wheel becomes the positive pole by means of a brush that smoothly contacts its surface. The negative electrode is fixed at a distance of about 0.3 mm from the wheel. In this small clearance between the fixed electrode and wheel surface, electrolysis occurs through the grinding fluid and an electric current. Figure 4.2 shows the schematic illustration of ELID single-side lap grinding.

The ELID single-side lap grinding system is comprised of grinding wheels, a pulse generator, a coolant box, a grinding fluid, and workpieces. In this experiment, the lapping wheels used were of the disk type, 250 mm in diameter and 55 mm in width. A chemical-solution-type grinding fluid was diluted by 2% with water.

4.1.3   EFFECTS OF ELID ON SINGLE-SIDE LAP GRINDING

The most important and significant aspect about ELID is the realization of mirror surface quality grinding. Figure 4.3 shows the surface roughness profiles by a #4000 CIB-D wheel using ELID and non-ELID.

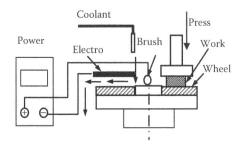

FIGURE 4.2   Schematic illustration of ELID single-side lap grinding.

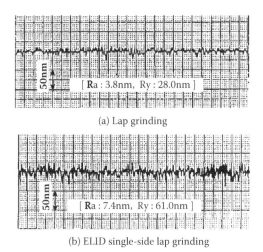

(a) Lap grinding

(b) ELID single-side lap grinding

FIGURE 4.3   Surface roughness profiles by a #4000 wheel using ELID and non-ELID. (a) Lap grinding. (b) ELID single-side lap grinding.

The workpiece is tungsten carbide. The ground surface roughness produced by ELID single-side lap grinding was 3.8 nm in Ra and the surface quality was excellent. On the other hand, with single-side lap grinding (without ELID), the finished surface roughness was 7.4 nm in Ra, and it was inferior to that by ELID single-side lap grinding.

The second significant effect of ELID is the continuity of the removal rate. Figure 4.4 shows the effects of ELID on the continuity of the removal rate of tungsten carbide. The wheel used was a #4000 CIB-D wheel, and the workpiece was tungsten carbide. ELID single-side lap grinding stabilized after 80 minutes without decrease in the removal rate under these operation conditions. Owing to the ELID

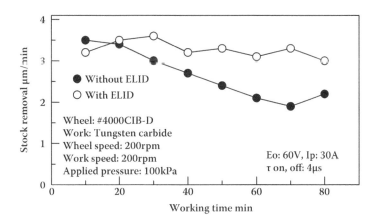

FIGURE 4.4   Effects of ELID on the continuity of the removal rate of tungsten carbide.

technique, grinding continuity was achieved and good surface finish was obtained without any irregularities with this system.

4.1.4   GRINDING CHARACTERISTICS OF MONOCRYSTALLINE SILICON AND TUNGSTEN CARBIDE

The experiments were conducted to study the effects of mesh number on the surface finish. Figure 4.5 shows the difference in the ground surface roughness and removal rate of monocrystalline silicon and tungsten carbide produced by different mesh numbers. For both materials, surface roughness values and removal rate decreased

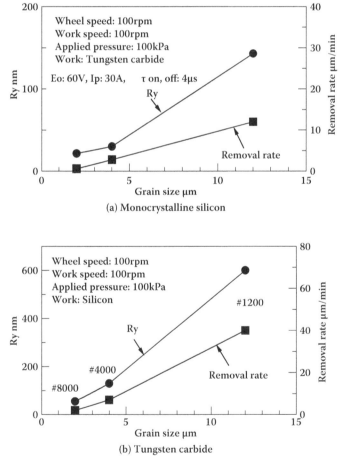

(a) Monocrystalline silicon

(b) Tungsten carbide

FIGURE 4.5   Difference in the ground surface roughness and removal rate. (a) Monocrystalline silicon. (b) Tungsten carbide.

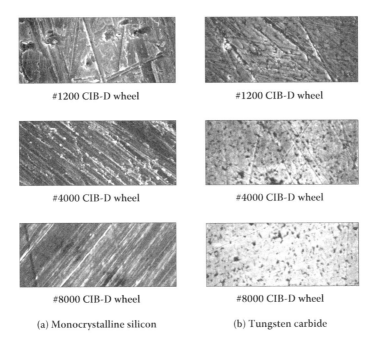

#1200 CIB-D wheel   #1200 CIB-D wheel

#4000 CIB-D wheel   #4000 CIB-D wheel

#8000 CIB-D wheel   #8000 CIB-D wheel

(a) Monocrystalline silicon   (b) Tungsten carbide

FIGURE 4.6   SEM photographs of the ground surface of monocrystalline silicon (a) and tungsten carbide (b) using the #1200 (top row), #4000 (middle row), and #8000 (bottom row) CIB-D wheels.

with the mesh number. The resultant surface roughness value of monocrystalline silicon was 7.4 nm Ra and that of tungsten carbide was 3.0 nm Ra using a #8000 CIB-D wheel. Mirror surface finish without any irregularities could be obtained. A difference in the removal rate of the two materials caused by machinability was seen.

Figure 4.6 shows the scanning electron microscope (SEM) photographs of the ground surface of monocrystalline silicon and tungsten carbide using the #1200, #4000, and #8000 wheels. In the case of monocrystalline silicon, the surface ground by the #1200 CIB-D wheel shows typical brittle fracture removal in the observation of SEM. On the surface ground by the #4000 CIB-D wheel, a small-scaled brittle fracture along grain path was obtained. The finished surface was not smooth, but the #8000 wheel produced the surface ground mostly in the ductile resume. In the case of this material, the brittle-ductile transition was performed using wheels over #8000 with ELID single-side lap grinding. For the tungsten carbide, the surface ground by the #1200 CIB-D wheel showed brittle fracture removal. The #4000 and #8000 CIB-D wheels produced the surface ground mostly in the ductile resume. In this case, brittle-ductile transition was performed using wheels over #4000 with ELID single-side lap grinding. Figure 4.7 shows the examples of monocrystalline silicon and tungsten carbide finished by ELID single-side lap grinding using a #8000 CIB-D wheel.

(a) Monocrystalline silicon   (b) Tungsten carbide

FIGURE 4.7   Samples of monocrystalline silicon and tungsten carbide using a #8000 CIB-D wheel. (a) Monocrystalline silicon. (b) Tungsten carbide.

4.1.5 SIMULTANEOUS GRINDING OF DIFFERENT MATERIALS

Simultaneous grinding is aimed at improving the ground surface roughness. Figure 4.8 shows the schematic illustration of simultaneous grinding. Material aiming for excellent surface is arranged inside, and the ring-shaped material, which is the harder material, is arranged on the outside. This complex workpiece was ground by ELID single-side lap grinding simultaneously. After grinding the ring-shaped material, the wheel surface was found to become smooth and the scatter of the tip of grain decreased by wear. As a result, excellent mirror surface finishing of material could be achieved.

In this case, a complex material composed of a monocrystalline silicon plate and tungsten carbide ring, as shown in Figure 4.9, was ground using this system. Figure 4.10 shows the experimental result of surface roughness and removal rate of monocrystalline silicon. When monocrystalline silicon was ground together with tungsten carbide, the surface roughness of monocrystalline silicon improved. Its removal rate was approximately the same as that of tungsten carbide, indicating that this method improves the surface roughness of monocrystalline silicon. The surface

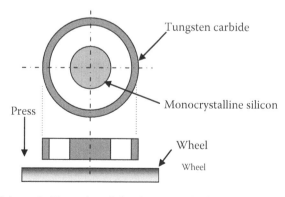

FIGURE 4.8   Schematic illustration of simultaneous grinding.

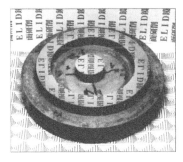

FIGURE 4.9   View of resultant mirror surface.

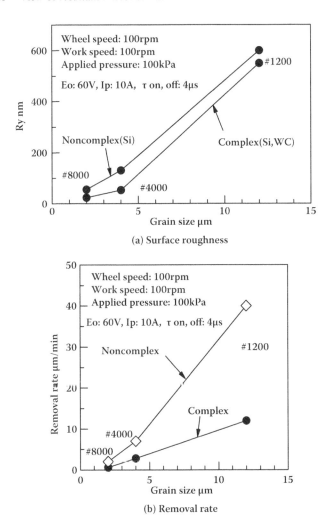

(a) Surface roughness

(b) Removal rate

FIGURE 4.10   Experimental results of surface roughness and removal rate of monocrystalline silicon. (a) Surface roughness. (b) Removal rate.

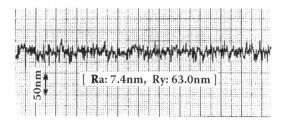

(a) Noncomplex grinding (monocrystalline silicon was ground alone)

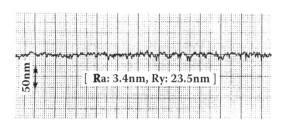

(b) Complex grinding (monocrystalline silicon was ground with tungsten carbide)

FIGURE 4.11 Surface roughness profiles of monocrystalline silicon using a #8000 CIB-D wheel. (a) Noncomplex grinding (monocrystalline silicon was ground alone). (b) Complex grinding (monocrystalline silicon was ground with tungsten carbide).

roughness improved from 63.0 nm in Ry to 23.5 nm in Ry using the #8000 CIB-D wheel, as shown in Figure 4.11. The surface quality produced by this system was excellent. The reason for this being the wheel surface is smoothed by the wear during the grinding of tungsten carbide.

Figure 4.12 shows the SEM photographs of the ground surface of monocrystalline silicon that was ground together with tungsten carbide. In this case, though the surface ground by the #1200 CIB-D wheel showed brittle fracture removal, the #4000 and #8000 CIB-D wheel produced surface ground mainly in the ductile resume. Brittle-ductile transition was performed using wheels over #4000. When silicon was therefore ground together with tungsten carbide, the surface qualities of monocrystalline silicon improved, in comparison with those by grinding monocrystalline silicon alone.

4.1.6 CONCLUSION

Features of the ELID single-side lap grinding method were introduced in this section. The surface quality produced by this system using different-sized grain wheels was excellent. The resultant surface roughness of monocrystalline silicon was 7.4 nm in Ra and that of tungsten carbide was 3.0 nm in Ra using the #8000 CIB-D wheel. Through the SEM observations, in the case of silicon, brittle-ductile transition was performed using wheels over #8000 with ELID single-side lap grinding. For tungsten carbide, that transition was performed using wheels over #4000. When silicon was

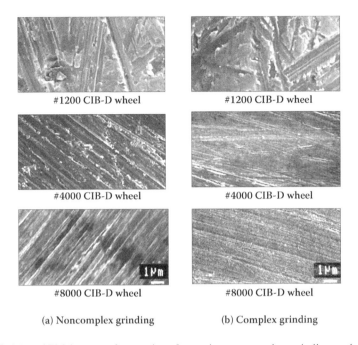

#1200 CIB-D wheel	#1200 CIB-D wheel
#4000 CIB-D wheel	#4000 CIB-D wheel
#8000 CIB-D wheel	#8000 CIB-D wheel
(a) Noncomplex grinding	(b) Complex grinding

FIGURE 4.12 SEM images of ground surface using noncomplex grinding and complex grinding. (a) Noncomplex grinding. (b) Complex grinding.

therefore ground together with tungsten carbide, the surface roughness of monocrystalline silicon improved, in comparison with those by grinding monocrystalline silicon alone. In this case, brittle-ductile transition was performed using wheels over #4000 with this system.

4.2 ELID DOUBLE-SIDE LAP GRINDING

4.2.1 INTRODUCTION

Demand for better surface roughness and higher accuracy of optical, electrical, and mechanical parts is growing with miniaturization and increasing high performance of high-tech devices such as computers and sensors. In addition, efficient machining of these parts is also essential for cost reduction. In general, the final finishing of these parts is performed by loose abrasive lapping and polishing, but these methods have such disadvantages as poor grinding efficiency, wastewater problem, mechanical damage, wear by scattering abrasive, dirty workplace, and difficulty in using wheels with different grains on the same machine. For these reasons, new alternative grinding methods to the finishing method using loose abrasives are sought. The authors have proposed a lap grinding method with special metallic-bonded wheels and ELID for the purpose of realizing smooth surfaces comparable to lapped or polished surfaces. As a part of this research and development, a new grinding machine, HICARION,[9] based on this grinding principle has been developed. This part introduces the feature of this machine and discusses its grinding performance.

FIGURE 4.13   External view of HICARION.

4.2.2   Features of HICARION

HICARION stands for *hi*ghly *c*ustomized and *a*dvanced *RI*EEN's invention *on* the machine. As it applies the ELID technique, finer grain wheels over #2000 can be used and mirror surface grinding can be achieved. In addition, two different grinding methods, single-side lap grinding and double-side lap grinding, can be switched by simple operations. Figure 4.13 shows the external view of HICARION.

4.2.3   ELID Double-Side Lap Grinding System

Figure 4.14 shows the schematic illustration of the ELID double-side lap grinding method.[10–14] In this case, same-size wheels were mounted on the top and bottom spindles, and the workpieces that were fixed by holder were sandwiched between the two wheels. This holder rotates together with the workpieces between the wheels to grind both sides efficiently. Figure 4.15 shows the close-up view of the double-sided grinding method. In this experiment, the wheels used were #4000 CIB-D, with a diameter of 250 rrm and width of 55 rrm. The workpieces used were tungsten carbide with a diameter of 30 mm. The speeds of both wheels were 100 rpm. The holder rotated at 30 rpm, and the applied pressure was 50 kPa.

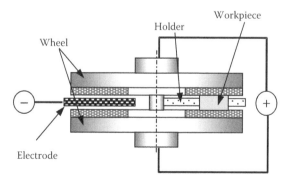

FIGURE 4.14   Schematic illustration of double-sided grinding.

FIGURE 4.15   Close-up view of the ELID double-side lap grinding method.

4.2.4 GRINDING CHARACTERISTICS

Figure 4.16 shows the relation between the grinding time and stock removal. The stock removal increased as the grinding time increased, and stable grinding was performed without clogging under these grinding conditions. Figure 4.17 shows the results of the surface roughness obtained. The surface roughnesses were good at both sides at 40 nm PV using this system. Figure 4.18 shows the mirror surface finish obtained.

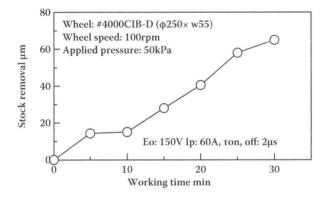

FIGURE 4.16   Relation between time and stock removal.

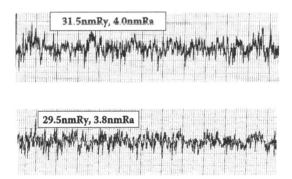

FIGURE 4.17   Surface roughness profiles of tungsten carbide.

FIGURE 4.18   Example of obtained mirror surface finish.

4.2.5   CONCLUSION

Features of the ELID double-side lap grinding method were introduced in this section. A stable grinding was performed without clogging, and good surface roughness could be obtained using HICARION. The experimental results show that mirror surface finish can be obtained without any irregularities using this system.

REFERENCES

1. N. Itoh and H. Ohmori, Grinding Characteristics of Hard and Brittle Materials by Fine Grain Lapping Wheels with ELID, *Journal of Materials Processing Technology* 62 (1996), 315–320.
2. N. Itoh and H. Ohmori, Finishing Characteristics of ELID-Lap Grinding Using Ultra Fine Grain Lapping Wheel, *International Journal of the Japan Society for Precision Engineering* 30, no. 4 (1996), 305–310.
3. N. Itoh, H. Ohmori, and B. P. Bandyopadhyay, Grinding Characteristics of Metal-Resin Bonded Wheel on ELID-Lap Grinding, *International Journal for Manufacturing Science and Production* 1, no. 1 (1997), 9–15.
4. N. Itoh, H. Ohmori, and B. P. Bangyopadhyay, Grinding Characteristics of Hard and Brittle Materials by ELID-Lap Grinding Using Fine Grain Wheels, *Materials and Manufacturing Process* 12, no. 6 (1997), 1037–1048.
5. N. Itoh, H. Ohmori, T. Kasai, and T. Karaki-Doy, Mirror Surface Grinding of ZrO_2 by Metal Resin Bonded Wheel with ELID, *Journal of the Japan Society for Precision Engineering* 31, no. 1 (1997), 55–56.
6. N. Itoh, H. Ohmori, T. Kasai, T. Karaki-Doy, and K. Horio, Characteristics of ELID Surface Grinding by Fine Abrasive Metal-Resin Bonded Wheel, *Journal of the Japan Society for Precision Engineering* 32, no. 4 (1998), 273–274.
7. N. Itoh, H. Ohmori, T. Kasai, and T. Karaki-Doy, Study of Precision Machining with Metal Resin Bond Wheel on ELID-Lap Grinding, *International Journal of Electric Machining* 3 (1998), 13–18.
8. N. Itoh, H. Ohmori, T. Kasai, and T. Karaki-Doy, Super Smooth Surface of X-Ray Mirrors by ELID-Lap Grinding and Metal-Resin Bonded Wheels, *Precision Science and Technology for Perfect Surfaces, JSPE* (1999), 103–108.

9. N. Itoh, H. Ohmori, T. Kasai, T. Karaki-Doy, and Y. Yamamoto, *Development of Double Sided Lapping Machine "HICARION" and Its Grinding Performance*, Proceedings of '99 AITC, CD-ROM (1999).

10. N. Itoh, H. Ohmori, T. Kasai, and T. Karaki-Doy, Development of Metal-Resin Bonded Wheel Using Fine Metal Powder and Its Grinding Performance, *Journal of the Japan Society for Precision Engineering* 33, no. 4 (1999), 335–336.

11. N. Itoh, H. Ohmori, T. Kasai, S. Moriyasu, S. Morita, and Y. Yamamoto, Mirror Surfaces Finishing on Double Sided Lapping Machine with ELID, *Proceedings of 1st USPEN* (1999), 266–269.

12. N. Itoh, H. Ohmori, T. Kasai, T. Karaki-Doy, T. Iino, and H. Yamada, Finishing Characteristics of Ceramics by ELID-Lap Grinding Using Fixed Fine Diamond Grains, *International Journal of New Diamond and Frontier Carbon Technology* 10, no. 1 (2000), 1–11.

13. N. Itoh, Y. Hasegawa, H. Ohmori, H. Yana, C. Uetake, T. Kasai, and T. Karaki-Doy, Development of Lapping Machine on Desk Top with ELID and Its Grinding Performance, *Advance in Abrasive Technology, JSAT* (2000), 411–416.

14. N. Itoh, Y. Hasegawa, H. Ohmori, H. Yana, C. Uetake, T. Kasai, and T. Karaki-Doy, Study on Mirror Surface Finishing Using Desk Top Lapping Machine with ELID, *Advance in Abrasive Technology, JSAT* (2001), 61–66.

5 ELID Honing

Weimin Lin and Hitoshi Ohmori

CONTENTS

5.1   ELID HONING PRINCIPLE

Electrolytic in-process dressing (ELID) grinding is a method in which electrolytic dressing is performed on metal-bond grinding wheels with fine abrasives by in-process to maintain sharpness of the grinding wheel at all times, thereby realizing high performance and high precision mirror surface grinding. As the ELID grinding method can be realized easily with the use of a conductive grinding wheel for ELID, power supply for ELID, electrodes, and grinding solution for ELID, it can be broadly applied to existing grinding methods. Already, ELID grinding is applied to various methods, and the required machining system is selected according to the required grinding surface properties; grinding surface accuracy; and grinding efficiency, ground surface quality, functionality, and so forth. Also taken into account are the required grinding wheel type, cutting method, contact method, and so forth.[1–3]

In the case of honing, due to the need to use honing wheels, there is no space to install electrodes on the opposite side of the grinding/honing wheel during the machining process. For this reason, interval ELID grinding must be performed.[4] Consequently, the appropriate setting of electrolytic conditions and machining conditions are key points for stable production. As shown in Figure 5.1, ELID honing was applied to the mirror finishing of the inner surface of parts made of brittle materials such as ceramics.[5] This ELID honing process consists first of (1) EDM truing of the metal grinding wheel, (2) initial dressing of the electrolytic electrode, and then (3) honing. Finishing is performed by repeating steps 2 and 3 according to the machining conditions and machinability of the workpiece material. In this way, the honing process is divided into two processes—dressing and grinding—thereby making it inappropriate for line productions of automobile engine blocks and so forth. It has mainly been used for parts made of conductive materials; however, measures are required to resolve electrolytic phenomena that occur during machining.

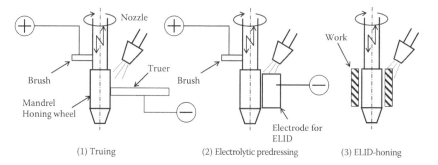

FIGURE 5.1   ELID-honing method.

5.2   ELID HONING SETUP

The ELID honing equipment is shown in Table 5.1. The base machine of the developed ELID honing machine is a vertical honing machine whose specifications are summarized in Table 5.2. This unit was incorporated with the ELID power supply, automatic electrode positioning device, and an exclusive program developed for it.

TABLE 5.1
Specifications of ELID Honing Experiment System

Machine type	Vertical honing machine
	ELID device: Electrode drive unit with EDM functions
Wheel	Metal-bond diamond wheel (30 × 3 mm)
	Mesh size: #325, #4000
	Concentration: 100
Power supply	ED-920
Fluid	AFG-M 2% dilution of water

TABLE 5.2
Specifications of Base Machine for ELID Honing Machine

Hole diameter (mm)	φ15-50/80 (110)
Hole length (mm)	110 (150, 200)
Spindle speed (rpm)	180–1300
Maximum reciprocating speed (m/min)	22
Stroke control system	NCS electrohydraulic servo system
Stone expansion system	Screw-wedge system (driven by stepping motor)
Sizing system	Sizematic/gaugematic

TABLE 5.3
Properties of Typical Ceramic Materials Used in ELID Honing

	Hardness (Hv) (Kg/mm^2)	Bending strength (Kg/mm^2)	Fracture Toughness (MN/m$^{3/2}$)	Poisson's Ratio
SiC	2400	60	3	0.16
ZrO$_2$	1200	120	10	0.31
WC	1650	220	12	0.21

Metal-bond diamond wheels composed of four each rectangular grindstone (honing stone) of #325 and #4000 with 100 diamond abrasive concentrations were used as the honing wheel. The grinding fluid used was AFG-M diluted 50 times with water.

5.3   ELID HONING APPLICATION TO HARD AND BRITTLE MATERIAL

The experiment workpiece of typical ceramic materials, whose properties are shown in Table 5.3, were ground in the experiments. The size of the workpiece was 30 mm in external diameter, 17.8 mm in internal diameter, and 30 mm in length.

After truing the wheels by EDM, ELID honing was performed. The #325 wheel was used for rough grinding and the #4000 wheel was used for fine grinding. The reliability of the honing operations was evaluated according to the stability of the spindle load.

Table 5.4 shows the ELID honing conditions for silicon carbide. Conditions were selected placing importance on reliability rather than performance. The cemented carbide was given a higher wheel speed than the other wheels.

TABLE 5.4
ELID Honing Conditions

Wheel mesh size	#325	#4000
Wheel revolution (rpm)	315	315
Feed rate (m/min)	5	3
Skip-time pass	1 (1st); 2 (2nd)	2 (1st); 4 (2nd)
Expansion (μm)	0.2 (1st); 0.2 (2nd)	0.2 (1st); 0.2 (2nd)
Total expansion (μm)	20 (1st); 40(2nd)	20 (1st); 40 (2nd)
Clean-up time (s)	10	30/10
Dwelling time (s)	0.3	0.3
ELID conditions	Open circuit voltage (*Eo*): 140 V; peak current (*Ip*): 3 A; on/off time (τ_{on}/τ_{off}): 5/2 μs	

Figure 5.2 shows the samples finished by the developed ELID honing machine. Using #4000 wheels, each workpiece achieved the surface roughness of 20 nm to 50 nm in Ra. The mirror surface can be found.

5.4   ELID HONING APPLICATION TO AUTOMOBILE PARTS

Based on the aforementioned information, experiments were conducted on ELID honing. To enable application to mass production, the same machining system and machining conditions as mass production were set. Prior to the experiments, the system used was remodeled. As shown in Figure 5.3, the electrodes were designed and arranged according to the system and the honing tool was installed so that electrolytic is carried out via the electrodes before honing. To prevent electrical corrosion, the system was equipped with a function that applies positive potential to the grinding wheel so that ELID cycle is carried out as required and adequately, and a poka-yoke function that detects and displays when ELID is being carried out. In preliminary experiments, actual ELID honing experiments were carried out while checking the ELID cycle. The engine block used in the experiments was the same as that mass produced. Continuous machining experiments were conducted on 100 engine blocks to evaluate machined surface roughness, roundness, and so forth, on the same evaluation system as that used for mass produced products. Comparisons were made between normal honing and ELID honing by randomly extracting and measuring the same number of mass-produced products.

Figure 5.4 shows the external view of the engine block machined by ELID honing. It looks completely the same as a block machined by the normal honing method, and shape accuracy was found to have improved. In addition, ground surface properties also improved. As shown in Figure 5.5, the finished roughness had improved to a certain extent compared to normal roughness (0.2 µm Ra), indicating the machined surface to be suited to the engine cylinder inner wall. Figure 5.6 shows the inconsistencies in roughness values (R_k) obtained for the same number of engine part samples machined by normal honing and ELID honing. Evidently, the application of ELID grinding reduces these inconsistencies in roughness values by more than half, contributing to quality stability and process management of the products. ELID honing was also found to shorten the honing process time from 70 to 80 seconds to about 40 seconds per cylinder. The use of conductive ELID solution also helps reduce environment load, and furthermore, the reduction of grinding time and grinding resistance realizes reduced power consumption, making this method an ecofriendly manufacturing process.

Based on these results, we applied the ELID honing technique to the production line of engine blocks. The following describes the results.[6–8]

1. Enhanced quality—The application of ELID honing was found to realize high grinding wheel performance stably. As shown in Figures 5.7 and 5.8, surface roughness improved and inconsistencies in surface roughness properties are reduced. Figure 5.8 shows the inconsistencies in these properties (R_{pk}, R_k, R_{vk}) of engine blocks finished by ELID honing and normal honing. The application of ELID honing was found to improve R_{pk} and

(a) Silicon carbide (SiC)

(b) Zirconium oxide (ZrO2)

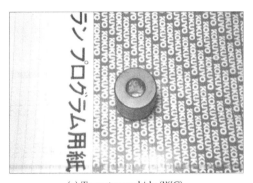

(c) Tungsten carbide (WC)

FIGURE 5.2 Samples finished by ELID honing. (a) Silicon carbide (SiC). (b) Zirconium oxide (ZrO_2). (c) Tungsten carbide (WC).

remarkably control inconsistencies in roughness properties. In addition, good sharpness is maintained, and excellent sliding surface properties with little plastic flow shown in Figure 5.9 were also obtained due to the use of grinding wheel with firm abrasives.

2. Shorted grinding cycle time—Mass production ELID honing is able to realize rough machining and finishing without having to change the

FIGURE 5.3   ELID honing machine.

FIGURE 5.4   External view of the engine block machined by ELID honing.

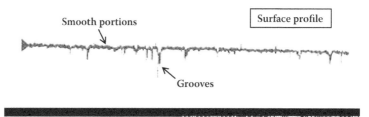

FIGURE 5.5   Surface roughness of ELID honing.

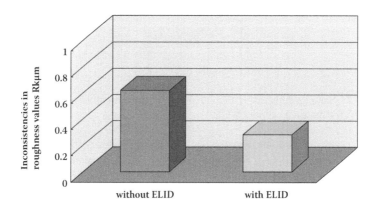

FIGURE 5.6   The inconsistencies in roughness values (R_k).

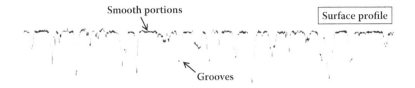

FIGURE 5.7 Surface roughness of mass production.

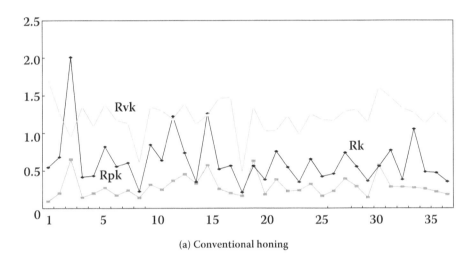

(a) Conventional honing

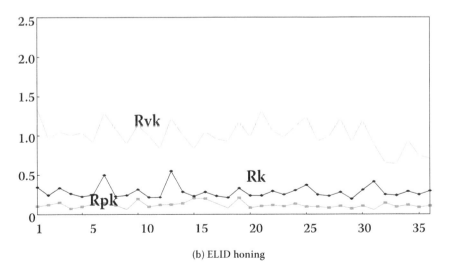

(b) ELID honing

FIGURE 5.8 Surface roughness of mass production. (a) Conventional honing. (b) ELID honing.

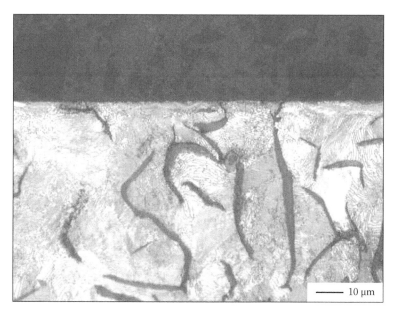

FIGURE 5.9   SEM image of engine section of mass production.

specifications of the honing wheels normally used. Moreover, the cycle time of both roughing and finishing could be shortened. Actual measurements showed that it was possible to shorten the actual machining time by 20% to 30% without changing the machining conditions in rough honing. By adjusting conditions, the actual machining time of one bore could be shortened by 40% to 50%, thus contributing to enhanced productivity.

3. Energy-saving effects—The power consumption of the main spindle of the honing system was measured on a production line used for ELID honing mass production. It was found that the power consumption per system was reduced by about 26% when the reduced time of the machining cycle and reduced grinding load were taken into consideration. In particular, as shown on the power waveform in Figure 5.10, the peak power does not indicate a protruded waveform at the start of machining, and the reduction rate of the power consumed is close to 30%.

4. Tool life increase effects—As a result of measuring grinding wheel wear of the honing wheel and predicting its lifespan, lifespan is predicted to extend by more than two times. Currently, the authors are continuing to observe this. The continuous protrusion effects of ELID are thought to help reduce forced self-sharpening actions, thus reducing wheel wear.

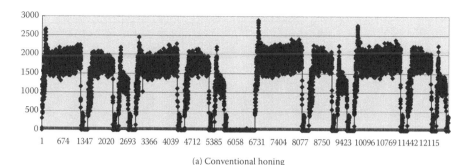

(a) Conventional honing

(b) ELID honing

FIGURE 5.10 Spindle power waveform of mass production. (a) Conventional honing. (b) ELID honing.

As can be estimated from the power waveform shown in Figure 5.10a, grinding resistance increases with the biting of the abrasives at the start of the roughing cycle in normal honing, resulting in the appearance of a peak power waveform. During the honing process, self-sharpening actions of the grinding/honing wheel occur, causing large wear of the wheel. As shown in Figure 5.10b, during ELID honing, the peak current waveform does not appear, and the current waveform of each cycle is more or less the same, which is thought to be due to the dressing effects of ELID grinding.

Figure 5.11 shows an scanning electron microscope (SEM) image of horning wheel surface. The abrasive had been found to increase due to the dressing effects of ELID grinding.

5. Costs reduction effects—As the application of ELID honing enables minimum remodeling with just the addition of electrodes and ELID power supply to existing facilities, investments in equipment can be reduced, and the stopping time of facilities can also be minimized. This helps cut new investment costs as well as increases the performance of production lines.

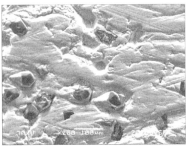

(a) Conventional honing

(b) ELID honing

FIGURE 5.11 SEM image of honing wheel surface. (a) Conventional honing. (b) ELID honing.

REFERENCES

1. H. Ohmori and T. Nakagawa, Analysis of Mirror Surface Generation of Hard and Brittle Materials by ELID (Electrolytic In-Process Dressing) Grinding with Superfine Grain Metallic Bond Wheels, *CIRP Annals Manufacturing Technology* 44, no. 1 (1995), 287–290.
2. H. Ohmori and T. Nakagawa, Utilization of Nonlinear Conditions in Precision Grinding with ELID (Electrolytic In-Process Dressing) for Fabrication of Hard Material Components, *CIRP Annals Manufacturing Technology*, 46, no. 1 (1997), 261.
3. J. Qian, H. Ohmori, and W. Lin, Internal Mirror Grinding with a Metal/Metal-Resin Bonded Abrasive Wheel, *International Journal of Machine Tools and Manufacture* 41, no. 2 (2001), 193–208.
4. C. Zhang, H. Ohmori, and W. Li, Small-Hole of Ceramic Material with Electrolytic Interval-Dressing (ELID-2) Grinding, *Journal of Materials Processing Technology* 105 (2000), 284–293.
5. H. Ohmori, T. Yamamoto, and I. D. Marinescu, Development of New Honing Machine with Electrolytic Interval Dressing Capability, *International Journal for Manufacturing Science and Technology* 1, no. 2 (1999), 75–79.

6. M. Shimano, Y. Yamamoto, J. Maruyama, W. Lin, and H. Ohmori, Development of ELID Honing Method (1st Report), *2007 JSAE Annual Congress Extended Summary* (2007), 65.

7. W. Lin, H. Ohmori, Y. Uehara, and T. Matsuzawa, Surface Quality Control Machining Technique with ELID Grinding, *2007 JSAE Annual Congress Extended Summary* (2007), 250.

8. M. Shimano, Y. Yamamoto, J. Maruyama, H. Ohmori, and W. Lin, Development of ELID Honing Method, *SUBARU Technical Review 2007,* 200–203.

6 ELID Free-Form Grinding

Takashi Matsuzawa and Hitoshi Ohmori

CONTENTS

6.1 FREEDOM PRINCIPLE

Presently, surface finishing of free-form die and mold is done by hand. However, improvement of work efficiency has realized automatic surface finishing in the recent years. Finishing of free-form die and mold by the ball-nose wheels was attempted in this study. Problems of change in form were however met due to the low rotation speed for the dead center point of the wheel was confirmed. ELID grinding was also applied for efficient machining. But conventional ELID grinding requires space for setting up the electrode to dress the grinding wheels outside the workpieces. To apply ELID grinding to the machining of certain types of grinding wheels with special shapes, interval dressing had to perform. A unique free-form machining method was developed to solve the problem. Experiments were also conducted on the grinding performance.

The method for free-form machining by ball-nose wheels was examined, but grinding by the dead-center point for the rotation at the axis of the wheels was difficult. A higher speed rotation of the rotational axis of the tool was required to solve the problem. This could be achieved by the free-form machining method because the half-sphere wheels have a suitable degree for the axis of the machining tool. Such wheels also widen the machining range. The machining method was named FREEDOM (free dominant machining) method. Figure 6.1 shows the principle.

6.2 FREEDOM TOOL

The grinding tool was prototyped based on the FREEDOM method. Its tool applies a mechanism for rotating the metal-bonded wheels at a rotation at an axis tilted 45 degrees to the machining tool from the center of the device. A disk-shaped wheel with an R cross-section was used instead of a semishape wheel. The desired speed

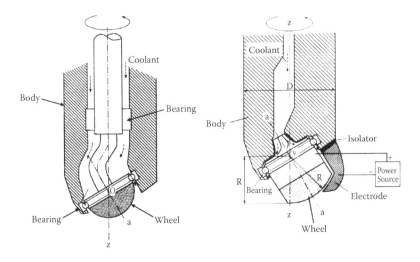

FIGURE 6.1 Principle of the FREEDOM method.

rotation was achieved at the dead point of the wheels using this system. To enable use of ELID grinding for this method, an electrode to dress the grinding wheels was set outside the workpieces. The tool was 50 mm across. The wheel was 30 mm across and 8 mm wide. The R cross-section was 4 mm. The wheels can rotate about 6000 rpm using a brushless D.C. motor and slope rotation drive system. This tool was named FREX-DO! Tool. Figure 6.2 shows a picture of the whole tool and the nose of the tool. Figure 6.3 shows the grinding state.

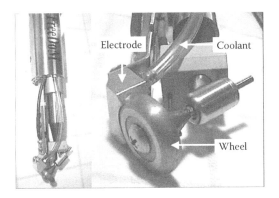

FIGURE 6.2 FREX-DO! Tool prototype.

FIGURE 6.3 Grinding state.

6.3 FREEDOM GRINDING

The grinding performance of FREX-DO! Tool was investigated. After predressing, ELID grinding was attempted for S35C and SUS420J2. The surface conditions of each wheel and surface roughness of the workpiece were investigated after grinding. Table 6.1 shows the experimental system.

Predressing of the wheels was carried out for 30 minutes. The grinding conditions were a wheel rotation speed of 1500 rpm, open voltage of 30 V, peak current of 10 A, and on/off time of 2 μs. The actual initial current values were small due to the small area of the electrode. The current values stabilized in 15 minutes and an oxide layer was formed on the surface wheel. After predressing, ELID grinding was attempted. Steady machining even at the dead center point of the tool was realized. The wheel rotation speed was comparatively fast at 500 m/min to 600 m/min. Better surface roughness was also obtained in spite of the curved surface wheels. But the tool vibrated during machining, resulting in decrease in the wheel rotation speed.

TABLE 6.1
Experimental System

Grinding machine	Vertical NC milling machine: FNC-105
Grinding wheel	Metal-bonded wheel/size: 30 × 8 × 8 mm
	cBN80, cBN200, cBN800, cBN1200, cBN4000
	SD80, SD200, SD800, SD4000
Power source	NX-ED ED910, ED920
Grinding fluid	NX-CL-CM2
Surface measurement	Surftest-701

TABLE 6.2
Grinding Conditions and Surface Roughness

	Straight					Curve
Wheel mesh	CBN80	CBN200	CBN800	CBN1200	CBN4000	CBN4000
Rotation speed (rpm)	6000	6000	6000	6000	6000	5000
Feeding (mm/min)	1000	500	500	200	150	100
Depth of cut (μm)	2	2	1	1	1	1
Pitch (mm)	0.2	0.2	0.2	0.2	0.2	0.32
Peak current (A)	10	10	10	10	10	10
Open voltage (V)	30	30	30	30	30	30
On/off time (μs)	2	2	2	2	2	2
Ra (μm)	0.384	0.344	0.159	0.148	0.128	0.085
Ry (μm)	2.150	1.888	1.389	0.990	0.636	0.315

FIGURE 6.4 Curve grinding sample.

TABLE 6.3
Grinding Condition to Spherical

Wheel mesh	SD80	SD200	SD800	SD4000
Wheel rotation speed (rpm)	5000	5000	4000	4000
Work rotation speed (rpm)	1000	1000	200	200
Feeding (mm/min)	10	6	1	0.4
Depth of cut (μm)	4	3	2	1
Peak current (A)	5	5	5	5
Open voltage (V)	30	30	30	30
On/off time (μs)	2	2	2	2

TABLE 6.4

Surface Roughness to Spherical

Wheel Mesh	Position	Ra (µm)	Ry (µm)
SD80	Center	0.900	5.650
SD200	Center	0.392	2.510
	Outside	0.834	5.240
SD800	Center	0.238	1.835
	Outside	0.246	2.120
SD4000	Center	0.010	0.060
	Outside	0.016	0.095

FIGURE 6.5 Spherical grinding sample.

The dead-center point of the wheel was also found to change shape. Steady machining and better surface roughness were obtained when no vibration was seen in the wheel rotation speed. Table 6.2 shows the grinding conditions and the surface roughness for straight and curve. Figure 6.4 shows the curve grinding sample. Table 6.3 shows the grinding condition for spherical. Figure 6.5 shows the spherical grinding sample. The obtained surface roughness is listed in Table 6.4.

Section III

Case Studies

7 Ultraprecision and Nanoprecision ELID Grinding

Kazutoshi Katahira and Hitoshi Ohmori

CONTENTS

7.1 ULTRAPRECISION ELID GRINDING OF SOLID IMMERSION MIRROR

The near-field optical recording technologies using a solid immersion lens (SIL; Figure 7.1) or solid immersion mirror (SIM; Figure 7.2) have been proposed as promising candidates to overcome the density limit of conventional optical recordings with far-field optics.

The recording density of 50 Gbit/square inch has been already attained with an SIL, which is the highest one in a single-layered recording. These devices are also suitable for optically assisted magnetic (OAM) recording technologies, avoiding the superparamagnetic problem in hard disk drive (HDD). However, a nonremovable type system is preferred to the near-field recording with an SIL or SIM because the air gap between their surface and recording layers should be reduced to less than 100 nm so the evanescent wave penetrates into the recording layers. In such kind of usage, the HDD is a main competitor and so the head size and weight, especially its height, should be reduced to those of the magnetic heads to make the volume density and transfer rate competitive. According to this standpoint, Fuji Xerox Co. proposed an optical head with a new type of a solid immersion mirror that consists of a hemiparaboloidal, reflecting surface, and incident and focused surfaces. The reflecting surface can concentrate an input-laser beam to the bottom surface of the mirror, and, at the same time, it can fold the laser beam toward the disk surface without an objective lens or folding mirror, which makes it possible to reduce the head height. In this study, we attempted the development of an electrolytic in-process dressing (ELID) grinding technique and on-machine measuring required mainly for paraboloidal surface.[1,2]

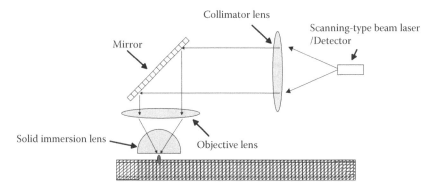

FIGURE 7.1   Solid immersion lens (SIL).

The configuration of the four-axes ultraprecision noncontact hydrostatic machine tool is shown in Figure 7.3. Aspherical and axis-asymmetric curved surfaces can be formed with this machine by single-point diamond cutting or a grinding process using the ELID grinding method. The machine has three linear axes (X, Y, Z) and one rotational axis (C). Those linear axes are controlled by an on-contact hydrostatic

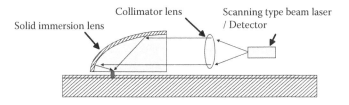

FIGURE 7.2   Solid immersion mirror (SIM).

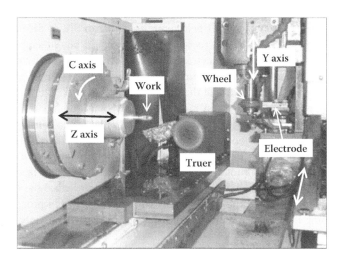

FIGURE 7.3   Construction four-axes ultraprecision machine tool.

TABLE 7.1

Specifications of Ultraprecision Machine Tool

Linear axis X, Y, Z	Moving range: X, 350 mm; Y, 100 mm; Z, 150 mm
	Scale resolution: 0.7 nm/8.7 nm
C axis	Precision air bearing
	Max speed: 1500 rpm
	Rotary encoder resolution: 1/4096 deg
Grinding spindle	Precision air bearing
	Rotary speed: 6000~20000 rpm
On-machine metrology probe	Measuring range
	Axis symmetric: Φ200mm
	Axis asymmetric: 250 mm × 50 mm

screw system and double laser linear scale to reach nanometer precision. The straightness and position repeatability of the linear axis for full stroke were better than 150 nm and 20 nm, respectively. In addition, step response of 2 nm was verified. The rotation spindle C is the ultraprecision pneumatic spindle that holds the workpiece. The maximum rotation speed of the ultrahigh precision grinding spindle mounted on a Y axis is 20,000 rpm. The profile accuracy can be measured by using the on-machine metrology probe mounted on the Y column without removing the workpiece from the machine tool. Specifications of the ultrahigh precision machine tool are listed in Table 7.1.

Figure 7.4 shows the manufacturing process of hemiparaboloidal SIM. Figure 7.5 shows the schematic of the tool path. The experimental conditions are summarized in Table 7.2. The metal-bonded grinding wheel to be used was trued mechanically and predressed using the ELID grinding method. After initial dressing, an experiment

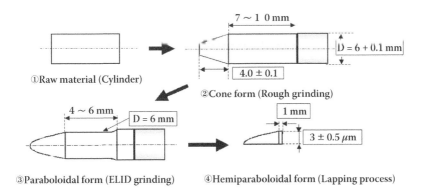

FIGURE 7.4 Manufacturing process of hemiparaboloidal SIM.

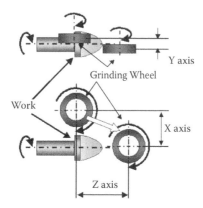

FIGURE 7.5   Tool path on grinding of hemiparaboloidal SIM.

was conducted. In this experiment, heavy flint glass of NbFD13 was used as the workpiece, and from #325 mesh to #3000 mesh cast-iron bonded diamond wheels were used as the grinding wheel. The height and width of the fabricated hemiparaboloidal SIM are 4.5 mm, and the total length is 5.5 mm. After grinding, the form was measured with the contact probe on the machine. The form data was fitted to the planned data with the least squares method, and the form error was calculated. The form error data was filtered with Fast Fourier Transform (FFT). The filtered form data and the compensated form data were generated by the computer. According to the compensated data, a new form was ground.

Figure 7.6 shows a solid immersion mirror after #3000 mesh ground. The finished surface was completely transparent and a mirror surface was successfully obtained. The profile accuracy of 0.6 μm PV in the range of the accuracy guarantee angle ±45 degrees was obtained (Figure 7.7). Figure 7.8 shows the surface roughness measurement after lapping process by atomic force microscopy (AFM). Although a fine surface on the circumference direction was obtained, a waving mark appeared on the optics axis direction surface. It was considered that this was perhaps caused by the

TABLE 7.2
Experimental Conditions

ELID conditions	Voltage (V)	60
	Maximum current (A)	10
	Pulse interval (μs)	2, 2
Grinding conditions	Grinding wheel	#325, #1200, #2000, #3000
	Wheel rotation speed	6000 rpm
	Work rotation speed	1000 rpm
	Feed rate (mm/min)	5
	Depth of cut (mm/pass)	0.01

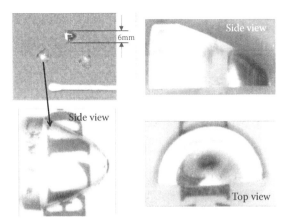

FIGURE 7.6 SIM after processing.

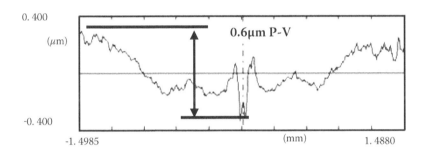

FIGURE 7.7 Profile accuracy of #3000 mesh.

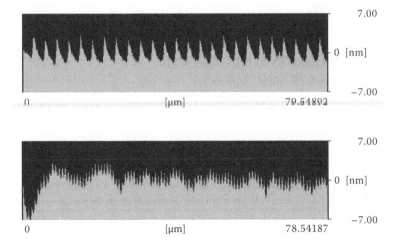

FIGURE 7.8 Surface roughness measurement by AFM.

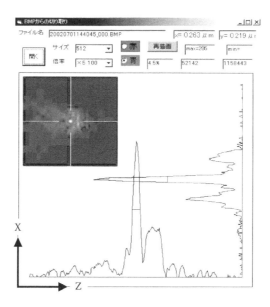

FIGURE 7.9   Condense result.

cutting depth in the fabrication processing. The concentration characteristics were studied using a blue laser beam ($\lambda = 405$ nm). The result is shown in Figure 7.9. The spot diameter of concentration in 0.26 μm on the circumference direction (X axis) and the spot diameter of concentration in 0.26 μm on the optics axis direction (Z axis) were achieved. Smaller spot diameter of concentration requires more ultraprecision surface roughness and higher profile accuracy. In the future we are going to optimize the experimental conditions while fabrication processing using a high-mesh-size grinding wheel.

7.2   ULTRAPRECISION ELID GRINDING OF GLASS-CERAMIC MIRROR

An efficient ultraprecision grinding process is important for research and developments, especially in optics such as astronomical observatory. It is also important in the field of industry to form ultraprecision dies for aspherical lenses, f-lenses, prisms, and so on. Also an ultraprecision metrological system combined with machine tools is very important for ultrahigh precision machining with high repeatability.

A four-axis ultraprecision noncontact hydrostatic driving system to meet the needs of three-dimensional nanolevel fabrication is introduced. A fabrication experiment of the paraboloidal mirror (50 mm in diameter) of Zerodur glass-ceramic material using an ELID grinding system was carried out to verify the system performance. The ground profile was measured by an on-machine metrology profile probe, then

FIGURE 7.10 Four-axis ultraprecision machine.

profile error was calculated to compensate and feedback to NC data during the grinding processing in order to achieve higher profile accuracy.

The configuration of the four-axis ultraprecision oil hydrostatic driving system is shown in Figure 7.10. Aspherical and axis-asymmetric curved surfaces can be formed with this machine by single diamond cutting and grinding processing using the ELID grinding method. The machine has three linear axes (X, Y, Z) and one rotational axis (C). The linear axes are controlled by an oil hydrostatic driving system and laser linear scale. The rotation spindle C is the main spindle that holds the workpiece. The maximum rotation speed of the ultrahigh precision grinding spindle mounted on the Y axis is 20,000 rpm. The profile accuracy can be measured without removing the workpiece from the chuck by using the on-machine metrology probe mounted on the Y column. The specifications of the ultrahigh precision machine tool are listed in Table 7.3.

The ultraprecision machine tool system was used to fabricate a concave paraboloidal mirror in this experiment. The material of the mirror used was Zerodur. The

TABLE 7.3
Specifications of Four-Axes Ultrahigh Precision Machine Tool

Linear axis X, Y, Z	Moving range: X, 350 mm; Y, 100 mm; Z, 150 mm
	Scale resolution: 0.7 nm/8.7 nm
C axis	Precision air bearing
	Max speed: 1500 rpm
	Rotary encoder resolution: 1/4096 deg
Grinding spindle	Precision air bearing
	Rotary speed: 6000~20000 rpm
On-machine metrology probe	Measuring range
	Axis symmetric: Φ200 mm
	Axis asymmetric: 250 mm × 50mm
	Resolution: 10 nm

TABLE 7.4

Experimental Conditions

	Finish grinding (#3000)
Wheel diameter	Φ75 mm
Tool feed	5 μm
Rotation speed	6000 rpm
Depth of cut	1 μm
ELID condition	60 V, 5A
	on: 2 μs; off: 2 μs

diameter of the mirror used is 50 mm and the maximum thickness is 50 mm. The grinding wheel rotated with high sending speed from the outer diameter to the center, and the workpiece was rotated during processing. The grinding wheel used was a cast-iron fiber-bonded diamond grinding wheel with grain size of #3000 and the outer diameter of 75 mm. Rotation speed of the grinding wheel was 6000 rpm. The depth of cutting and sending speed can be changed at any time if necessary. The ELID power supply used was ED920. Table 7.4 shows the ELID grinding conditions, and Figure 7.11 shows the state of paraboloidal mirror processing. In addition, compensation feedback using an on-machine metrology probe was applied during paraboloidal mirror fabrication.

Figure 7.12 shows a form error of grinding wheel. The ground profile was measured by an on-machine metrology profile probe, then the profile error was calculated to compensate and feedback to NC data during each grinding processing to achieve higher profile accuracy. Figure 7.13 shows the results of the surface roughness measured by the NEW VIEW (Zygo). The finished surface was completely transparent and mirror-like. The surface roughness of 220 nm in Ry, and 42 nm in RMS at the center was obtained. The processing could be completed within about 30 hours in total, which could be considered to be an efficient and stable process as an ultrahigh precision fabrication.

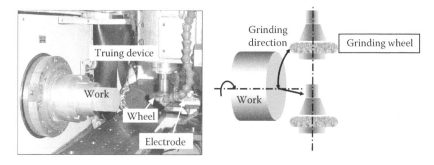

FIGURE 7.11   State of paraboloidal lens processing.

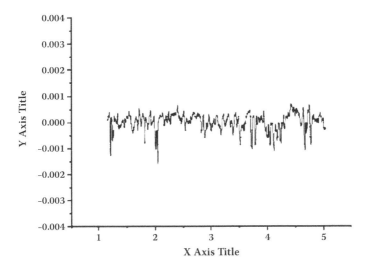

FIGURE 7.12 Form error of grinding wheel.

7.3 LARGE-SCALE ULTRAPRECISION AND NANOPRECISION ELID GRINDING

To grind metal mold steel effectively, the concept design of the large ultraprecision mirror surface machining system was proposed as shown in Figure 7.14. The machine has three linear axes, which can be controlled at a feeding resolution of 10 nm under full-closed feedback. The step response of Z-axis movement by feed resolution of 10 nm is shown in Figure 7.15 as measured using Microsense. A unique double hydrostatic guideway is used for the three axes sliding the machine. That is composed of a main table and subtable (see Figure 7.16). It makes sliding straight

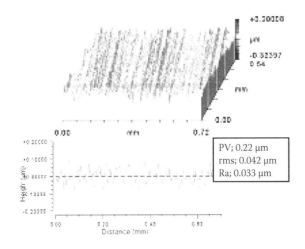

FIGURE 7.13 Surface roughness.

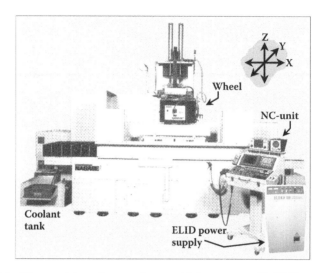

FIGURE 7.14   Ultraprecision mirror surface machining system with ELID.

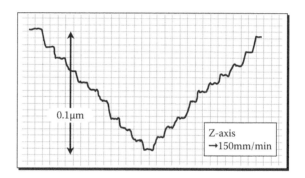

FIGURE 7.15   Step response of Z-axis movement by feed resolution of 10 nm.

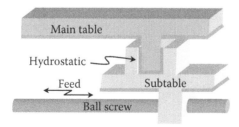

FIGURE 7.16   Schematic illustration of double hydrostatic guideway.

TABLE 7.5

Specifications of Developed Grinding System

Linear Axis: Double Hydrostatic Guideways

X axis: Stroke = 1400 mm, feed rate = 1–9000 mm/min

Y axis: Stroke = 550 mm, feed rate = 1–1000 mm/min

Z axis: Stroke = 500 mm, feed rate = 1–1000 mm/min

(Feeding resolution: 10 nm for each axis)

Rotational Axis: Hydrostatic Bearing

Grinding wheel spindle: 1800 rpm in maximum

Grinding wheel motor: 11 kw 4P

ELID Capacity

90 V (open voltage), 20 A (mean current), 40 A (peak current)

Duty ratio: 80% in maximum, square wave pulse generator

very smooth, and provides high accuracy, high damping, and dynamic stiffness. Hydrostatic bearing is also used for the grinding wheel spindle. The specifications of the developed machine are shown in Table 7.5. The maximum workpiece size is 1200 × 500 mm along the X and Y axes. The X axis can be driven at the 9 m/min in maximum feed speed.[3,4]

The developed machine is mainly composed of the high stiffness machine frame, NC controller, ELID system, and supporting devices such as oil tanks and temperature control units. Figure 7.17 shows the external view of the grinding wheel with the ELID electrodes. The chemical-solution-type coolant is supplied to the grinding

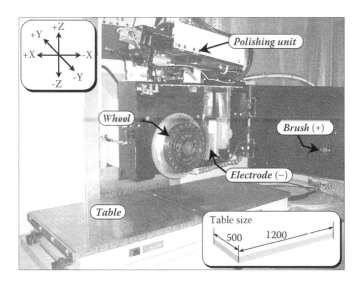

FIGURE 7.17 External view of the grinding wheel with the ELID electrodes.

wheel for ELID, and for cooling, cleaning, and lubricating the grinding points. The electrode can be adjusted by using NC to a specific gap.

The ELID power supply has a capacity open-circuit voltage of 90 V and a peak current of 20 A, generating square pulse undone of 1 to 10 msec during the on–off time at a maximum duty ratio of 80%. The power supply can be switched automatically by the M code in the NC program. The NC unit used can be connected with a personal computer through high-speed data transfer interface, and can display information including position data, controlling parameters, and NC programs on the personal computer console.

ELID grinding of a metal mold steel was conducted. The workpiece was a metal mold steel (SKD11) measuring 250 mm in length, 200 mm in width, and 50 mm in height. Cast iron–cobalt hybrid bonded #325, #1200, and #4000 diamond wheels were trued on the machine by the plasma discharge method. The truing wheel used was the metallic bond diamond type. Table 7.6 summarizes the grinding conditions. Relatively high efficiency could be achieved under these conditions to profile and finish the mirror surface by ELID. The finished surface roughness was approximately 0.057 microns in Ry, and 0.007 microns in Ra by the #4000 cast iron–cobalt hybrid diamond wheel. Figure 7.18 shows an overview and measured straightness of the workpiece after finishing. A very accurate surface with straightness of 0.25 microns per 250 mm length was attained.

As a further application of the ELID process, we fabricated large-scale optics intended for the development of physical elements. In particular, a large-scale telescope mirror, large-scale specially designed Schmidt plates, a parabolic mirror for use in Raman spectroscopy, forming dies for a Walter mirror, and synchrotron mirror were successfully developed. Figure 7.19 shows examples of large-scale optics produced by ELID. The finished quality has a surface roughness Ra value that is smoother than 3 nm in general. So when these optics are used for short wavelengths, they must be further finished or polished to have a surface roughness Ra value that is smooth in the subnanometer range.

TABLE 7.6
ELID Grinding Conditions (#4000)

Grinding Conditions

Wheel speed: 4 m/min

Depth of cut: 1.0 μm (coarse), 0.5 μm (finish)

Feed pitch (Y axis): 1.0 mm (coarse), 0.5 mm (finish)

Table feed rate (X axis): 4 m/min

ELID Conditions

Open voltage (Eo): 90 V, peak current (Ip): 20 A

Pulse timing (on/off): 4 μs

Pulse wave: Rectangle

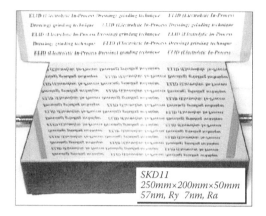

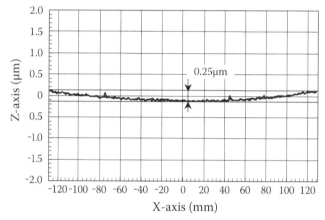

FIGURE 7.18 Overview and measured straightness of workpiece finished with ELID grinding.

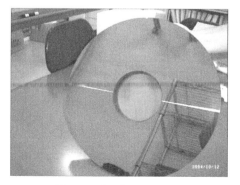

a) Large-scale telescope mirror

FIGURE 7.19 Examples of large optics produced by ELID. (a) Large-scale telescope mirror. (b) Large-scale specially designed Schmidt plates. (c) Parabolic mirror for use in Raman spectroscopy. (d) Walter mirror-forming dies. (e) Synchrotron mirror.

b) Large-scale specific designed Schmidt plates

c) Parabolic mirror for use in Raman spectroscopy

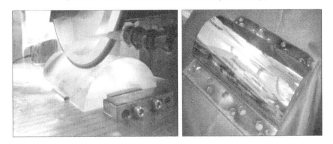

d) Walter mirror forming dies

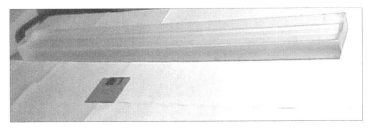

e) Synchrotron mirror

FIGURE 7.19   (Continued).

REFERENCES

1. Y. Uehara, H. Ohmori, Y. Yamagata, S. Moriyasu, T. Suzuki, K. Ueyanagi, Y. Adachi, T. Suzuki, and K. Wakabayashi, Grinding Characteristics of Solid Immersion Mirror with ELID Grinding Method, *Key Engineering Materials* 238–239 (2003), 83–88.
2. T. Suzuki et al., Ultraprecision Fabrication of Glass Ceramic Aspherical Mirror by ELID-Grinding with Nano-Level Positioning Hydrostatic Driving Machine, *Proceedings of the 3rd International Conference and 4th General Meeting of the European Society for Precision Engineering and Nanotechnology* 2 (2002), 795–798.
3. H. Ohmori, K. Katahira, M. Anzai, A. Makinouchi, Y. Yamagata, S. Moriyasu, and W. Lin, Mirror Surface Grinding Characteristics by Ultraprecision Multi-Axis Mirror Surface Machining System, *Journal of JSAT* 45, no. 2 (2001), 85–90. (In Japanese)
4. K. Katahira, H. Ohmori, M. Anzai, A. Makinouchi, S. Moriyasu, Y. Yamagata, and W. Lin, Grinding Characteristics of Large Ultraprecision Mirror Surface Grinding System with ELID, *Advances in Abrasive Technology* (2000), 125–128.

8 Desktop and Micro-ELID Grinding

Yoshihiro Uehara and Hitoshi Ohmori

CONTENTS

8.1 BACKGROUND OF DEVELOPMENT OF DESKTOP MACHINES

Micromechanical fabrication is the process of manufacturing ultraprecision microparts by microgrinding, micromilling, microcutting, microlapping, micropolishing, and micromolding. We established microfabrication tools using desktop machines mounted with an electrolytic in-process dressing (ELID) system,[1–2] and conducted investigations on its performances and applications, to answer to demands of new microfabrication technologies. The miniaturization and also the reduction of the weight of portable audio and visual equipment, palmtop computers, and cell phones are being urgently pursued and have been achieved recently. This has led to a growing demand for the miniaturization of even highly functional electronic devices, optical devices, and mechanical components that compose those apparatus, as well as high-dimensional accuracy of these parts. The dimensional accuracy requirement is increasingly difficult to obtain, as it is influenced by surface roughness and waviness. In particular, for optical parts, profile accuracy and surface roughness are demanded at the same time within the nanometer level. The improvement of surface accuracy for mechanical components is also desired. The proposed new microfabrication method and system are expected to satisfy the micromanufacturing requirements of these parts. The concept is summarized as a "micro-workshop" (as shown in Figure 8.1), which is a production shop composed of the desktop machines and units. This section describes the advanced desktop micromanufacturing system mainly applying ELID grinding. [3–12]

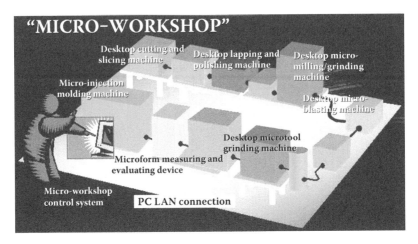

FIGURE 8.1   Concept of "micro-workshop."

8.2   CONCEPT OF DESKTOP FABRICATION SYSTEMS

The developed desktop fabrication system employed with the ELID grinding process is composed of the following machines and units:

- Desktop slicing machine
- Desktop lapping machine
- Desktop multiaxes grinding/cutting machine
- Desktop microtool grinding machine
- Micro-ELID (MQL/electrodeless) and truing devices
- (On-machine) measuring function/unit

These machines and units can work individually and work harmoniously. Each desktop machine can mount an ELID grinding system to enable high accuracy machining for hard materials. Because grinding resistance can be minimized by the ELID grinding performance and the profile of a grinding wheel can be kept for a long time by the high toughness of its metallic bond, a good surface roughness is efficiently obtained with high mesh sizes. The features of the developed desktop machines are as follows, also as shown in Figure 8.2.

- Compact body as desktop
- Lightweight
- Easy operations with personal computer (PC)-based NC controller
- Equipped with ELID grinding unit for mirror surface machining
- 100 V AC input power
- On-machine measuring capability

The authors therefore developed the desktop machine series, and through the use of these machines wider ranges of materials microfabrications for semiconductors,

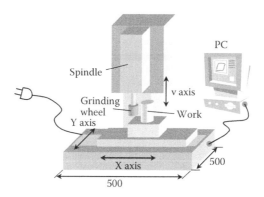

FIGURE 8.2   Example of desktop machine concept design.

optical elements, ceramic/carbide-alloy tools, and bioimplant materials were carried out for the first time on desktop. For these purposes, the systems must employ the ELID grinding process and the related mechanical processing to attain

• Hard and brittle material machining;
• Mirror surface finishing at the nanolevel;
• Low grinding resistance; and
• Low tool wear, high-/ultraprecision machining.

8.3   DEVELOPMENT OF DESKTOP FABRICATION SYSTEMS

8.3.1   DESKTOP SURFACE GRINDING/SLICING MACHINE

Figure 8.3 shows the developed desktop-type slicing machine. This machine is a desktop high-speed reciprocating feed type ELID surface grinding and slicing machine, which can be mounted with a thin-edge grinding wheel. Figure 8.4 shows

FIGURE 8.3   View of developed desktop slicing machine.

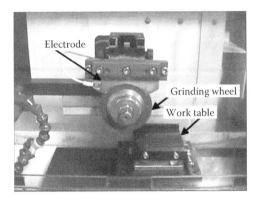

FIGURE 8.4   Close-up view of ELID unit for desktop slicing machine.

the equipped ELID unit. Figure 8.5 shows the results of slicing with ELID. The remarkable transparency on the sliced (grooved inside) surface can be shown. A mirror-like surface can be obtained just by slicing, contributing to reduced labor and time in the following lapping process.

FIGURE 8.5   Result of desktop slicing (grooving) for optical glass.

FIGURE 8.6   View of desktop ELID lap grinding machine.

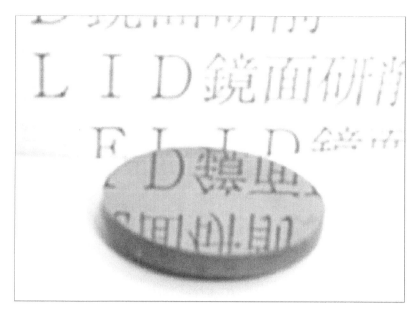

FIGURE 8.7 Result of ELID lap grinding for tungsten carbide.

8.3.2 DESKTOP LAP GRINDING MACHINE

Figure 8.6 shows the desktop ELID lap grinding machine, and Figure 8.7 shows results of ELID lap grinding on this machine. Very smooth surfaces with nanometer level are obtainable. Through high grinding accuracy, high grinding efficiency, and outstanding cost effectiveness, this machine is suited for fixed abrasive lapping of micro-optical, electronic, and mechanical parts.

8.3.3 DESKTOP MULTIAXES GRINDING/MILLING/CUTTING MACHINES

Figure 8.8 shows the desktop three-axis grinding/milling machine. Figure 8.9 shows the ELID grinding setup on this machine. Three-dimensional (3D) milling

FIGURE 8.8 View of desktop three-axis machine.

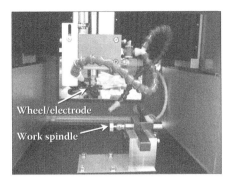

FIGURE 8.9   Example of ELID grinding setup.

can be performed by three-axes simultaneous control. A 3D profile measurement function is also possible on the same machine by replacing the toolhead with a measuring head.

Figure 8.10 shows the desktop four-axis machine. Figure 8.11 shows examples of ELID grinding on this machine. Consequently, curving, engraving, grooving, and profiling can be carried out by cutting or milling, and after these procedures, ELID grinding for mirror surface finish can be applied on the same system without detachment of the workpiece. Applications to microaspheric lens or molds and microparts fabrication can be looked forward to. The rotary index table, serving as the fourth axis, can be continuously rotated and angularly controlled at the same time, and thus used for axis-asymmetrical machining. A mirror surface can also be obtained by cutting for ductile materials. Figure 8.12 shows the mirror finished sample (oxygen free copper) by cutting.

8.3.4   DESKTOP MICROTOOL GRINDING MACHINES

To realize further desktop micromachining features, the development of microtools is indispensable. Figure 8.13 shows the schematic illustration of the first developed microtool grinding principle. Figure 8.14 shows the developed machine with the ELID grinding system. Figure 8.15 shows examples of microtool grinding results

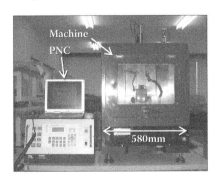

FIGURE 8.10   View of desktop four-axis machine.

(a) Tungsten carbide concave lens mold

(b) Optical glass aspheric lens

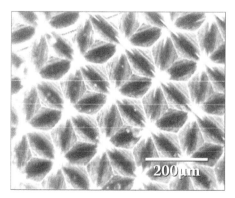

(c) Tungsten carbide reflector die

FIGURE 8.11 Examples by ELID grinding. (a) Tungsten carbide concave lens mold. (b) Optical glass aspheric lens. (c) Tungsten carbide reflector die.

(a) Turning operation

(b) Mirror surface obtained

FIGURE 8.12 Mirror surface by cutting (turning operation). (a) Turning operation. (b) Mirror surface obtained.

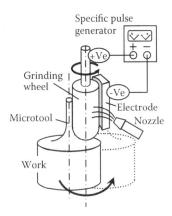

FIGURE 8.13 Microtool grinding principle.

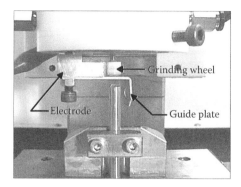

FIGURE 8.14 Close-up view of grinding machine with ELID unit.

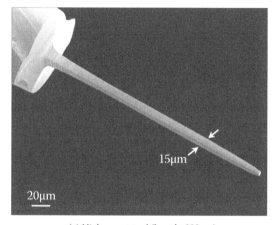

(a) High aspect tool (length: 300μm)

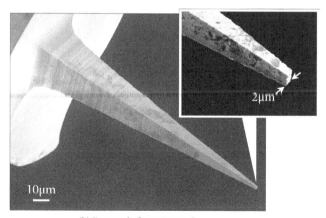

(b) Extremely fine microtool

FIGURE 8.15 Grinding results of microtool. (a) High aspect tool (length: 300 μm). (b) Extremely fine microtool.

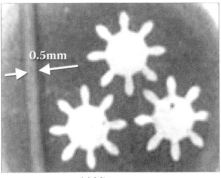

(a) Microgears

(b) Microstamped hole

FIGURE 8.16   Machining examples by microtools. (a) Microgears. (b) Microstamped hole.

with ELID. High aspect and sharp microtools could successfully be fabricated without crack origin by the mirror finish grinding feature. A pyramidal micro-tool with a top square of 2 μm × 2 μm was successfully achieved (Figure 8.15b). Figure 8.16 shows examples of microgears and microstamping by the fabricated tools.

The second developed microtool grinding machine has the new principle to grind microtools as shown in Figure 8.17. This method uses two grinding wheels that grind

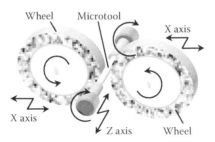

FIGURE 8.17   Microtool grinding principle with two wheels.

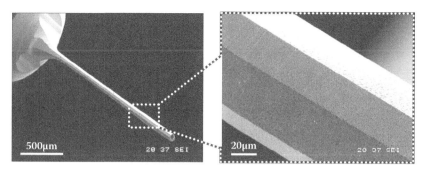

FIGURE 8.18 Examples of microtools ground by two wheels.

a workpiece for a microtool at the same time. To support deformation of a workpiece of a small diameter by each grinding wheel in-feed, an extremely high aspect ratio microtool grinding can be achieved. Figure 8.18 shows an example of an octagonal microtool obtained by this new method, which has about 30 times efficiency of the aforementioned method. Square, hexagonal, and octagonal microtools can be fabricated by indexing the workpiece.

REFERENCES

1. H. Ohmori et al., Mirror Surface Grinding of Silicon Wafer with Electrolytic In-Process Dressing, *CIRP Annals Manufacturing Technology* 39, no. 1 (1990), 329–332.
2. H. Ohmori et al., Analysis of Mirror Surface Generation of Hard and Brittle Materials by ELID (Electrolytic In-Process Dressing) Grinding with Superfine Grain Metallic Bond Wheels, *CIRP Annals Manufacturing Technology* 44, no. 1 (1995), 287–290.
3. H. Ohmori, State-of-the-Art on "MICRO-WORK-SHOP," *Proceedings of the 1st Japan-Korea Joint Symposium on Micro-Fabrication, RIKEN Microfabrication Symposium*, 5 (1999), 56–60. (In Japanese)
4. Y. Uehara et al., Development of "Micro-Workshop" by RIKEN-Microfabrication Using Desktop Machine with ELID System, *Proceedings of the 2nd Korea-Japan Joint Symposium on Micro-Fabrication*, KITECH (2001), 97–106.
5. Y. Uehara et al., Development of Small Tool by Micro-Fabrication System Applying ELID Grinding Technique, *Initiatives of Precision Engineering at the Beginning of Millennium* (2002), 491–495.
6. N. Itoh et al., Flattening of Micro-Functional Parts by ELID Lap Grinding, *10th International Conference on Production Engineering* (2001), 486–490.
7. H. Ohmori et al., Ultraprecision Micro-Grinding of Germanium Immersion Grating Element for Mid-infrared Super Dispersion Spectrograph, *Annals of the CIRP*, 50, 1 (2001), 221–224.
8. H. Ohmori et al., ELID-Grinding of Microtool and Applications to Fabrication of Microcomponents, *International Journal of Nano Technology* 41, no. 2 (2002), 193–204.
9. H. Ohmori et al., Improvement of Corrosion Resistance in Metallic Bio-Materials Using a New Electrical Grinding Technique, *CIRP Annals Manufacturing Technology* 51, no. 1 (2002), 491–494.

10. H. Ohmori et al., Improvement of Mechanical Strength of Micro Tools by Controlling Surface Characteristics, *CIRP Annals Manufacturing Technology* 52, no. 1 (2003), 467–470.

11. H. Ohmori et al., Investigation on Color-Finishing Process Conditions for Titanium Alloy Applying a New Electrical Grinding, *CIRP Annals Manufacturing Technology* 53, no. 1 (2004), 455–458.

12. Y. Uehara et al., Micro-Fabrication Techniques and Its Applications with Desk-Top Type Machining, *Proceedings of the 5th Japan-Korea Joint Symposium on Micro-Fabrication*, RIKEN (2005), 88–108.

9 ELID Grinding Applications

Hitoshi Ohmori, Ioan D. Marinescu,
Kazutoshi Katahira, Yutaka Watanabe,
Hiroshi Kasuga, Jun Komotori, Shaohui Yin,
Masayoshi Mizutani, and Tetsuya Naruse

CONTENTS

9.1   SEMICONDUCTORS

9.1.1   ELID GRINDING SYSTEMS OF SILICON WAFERS

A wafer is a thin slice of semiconductor material, such as a silicon crystal, used to fabricate integrated circuits and other microdevices. Electrolytic in-process dressing (ELID) grinding technology is suitable for achieving mirror-quality surface finishing of silicon wafers. Figure 9.1 shows a basic ELID rotary grinding system with a cup-type grinding wheel. Figure 9.2 shows an image of the grinding wheel surface after stable grinding. The white regions in the image are the fine diamond abrasives (3 μm). The wheel surface is covered with an insulating layer. Figure 9.3 shows the obtained surface roughness. A surface roughness of 20 nm in Rz was achieved using a #4000 grinding wheel. Figure 9.4 shows an example of subsurface damage. It is possible to reduce the amount of subsurface damage (less than 0.4 μm) using a #8000 grinding wheel. A higher quality surface is generally obtained using a finer grinding wheel. Figure 9.5a shows an image of a finished silicon wafer and Figure 9.5b shows its surface flatness.

Table 9.1 gives typical specifications for a practically used ELID grinding machine for semiconductor wafers. Reasonably stable ELID grinding is achieved due to using a built-in air hydrostatic motor spindle for the work and wheel spindles. The silicon wafer is fixed on the work spindle by a vacuum chuck table. Figure 9.6 shows an

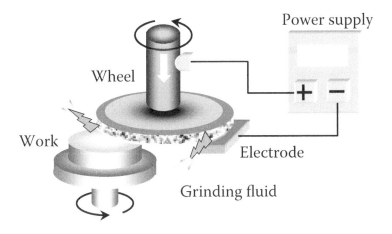

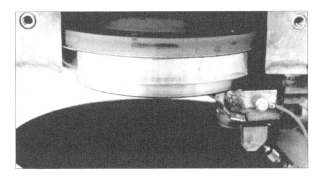

FIGURE 9.1   Basic ELID rotary grinding machine with cup-type wheel.

FIGURE 9.2   Image of grinding wheel surface after stable grinding process.

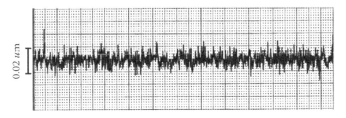

FIGURE 9.3   Surface roughness (#4000).

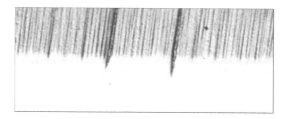

FIGURE 9.4 Subsurface damage.

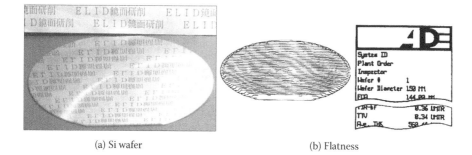

(a) Si wafer   (b) Flatness

FIGURE 9.5 ELID ground Si wafer. (a) Si wafer. (b) Flatness.

overview of the developed ELID grinding machine, which has precise motion control with a depth of cut of 0.1 μm.

Figure 9.7 shows an ELID grinding machine with a different machine configuration that has straight grinding wheels. It is designed to finish large silicon wafers with diameters of 300 mm or greater. Figure 9.8 shows an overview of this machine. Table 9.2 shows the basic grinding test conditions and the obtained surface roughness

TABLE 9.1
Specifications of ELID Grinding Machine for the Semiconductor Wafer

Work size	Φ8 inch
Grinding method	In-feed type grinding with cup-type wheel
Wheel spindle	Air hydrostatic built in motor spindle (AC3.7 kW4P, 300~3600 rpm)
Z-axis stroke	80 mm
Feed	0.001~1000 mm/min
Positioning resolution	0.0001 mm
Wheel size	φ350 × W3 mm
Work spindle	Air hydrostatic built-in motor spindle (1.5 kWDD motor, 300~1800 rpm)
X-axis stroke	220 mm
Feed	50~200 mm/min

FIGURE 9.6 Practically used ELID grinding machine for Si wafer.

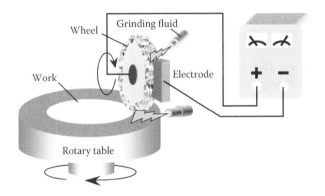

FIGURE 9.7 ELID grinding for large-scale Si wafer.

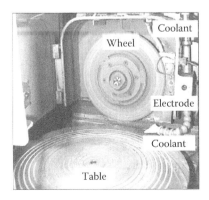

FIGURE 9.8 ELID grinding machine with straight-type wheel.

TABLE 9.2

Basic Grinding Test Conditions and Results

Grinding wheel	Straight-type metal bond grinding wheel (#6000)
Wheel rotation speed	1350 m/min
Work rotation speed	283 m/min
Feed	270 mm/min
Depth of cut	1 μm
Obtained surface roughness	0.0168 μm Ra, 0.13 μm Rz

results. Figure 9.9 shows an image of a φ300 mm Si wafer finished by the straight grinding method. Furthermore, a large-scale ultraprecision rotary ELID grinding system (Figure 9.10) has been developed for future applications.

A chip-scale package (CSP) has been developed as a new integrated-circuit chip carrier. In order to qualify as chip scale, the package must have an area no greater than 1.2 times that of the die and it must be a single-die, direct-surface-mountable package. Therefore, the fabricated wafer must be as thin as possible. The aforementioned rotary ELID grinding system with a cup wheel has been used for grinding the backside of Si wafers, as shown in Figure 9.11a. Figure 9.11b shows an image of the thickness evaluation after ELID grinding. A thin Si wafer (50 μm or less in thickness) was produced using the ELID system.

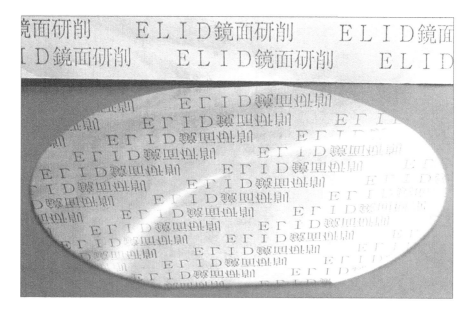

FIGURE 9.9   ELID ground φ300 mm Si wafer.

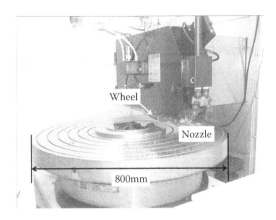

FIGURE 9.10 Large-scale ultraprecision ELID rotary grinding machine.

9.1.2 ELID Lapping of Monocrystalline Silicon

This part introduces the grinding characteristics of monocrystalline silicon by ELID single-side lap grinding using a #3000000 metal-resinoid hybrid bonded diamond (MRB-D) wheel. The lapping wheel used is composed of copper and resin, and has fine grain diamond grains of 5 nm in diameter (a #3000000 MRB-D wheel). The concentration is 75 and the ratio of copper to resin is 7:3. After examining the electrical behavior of the predressing of the metal-resin bonded wheel, ELID lap grinding was started and the grinding characteristics of the #3000000 MRB-D wheel were investigated.[1–3]

Important and significant effects of ELID on grinding performance are the realization of mirror-surface grinding and the continuity of the removal rate.

(a) Backside grinding of Si wafer (b) Evaluation of thickness after ELID grinding

FIGURE 9.11 ELID ground Si wafer (50 μm or less in thickness). (a) Buck side grinding of Si wafer. (b) Evaluation of thickness after ELID grinding.

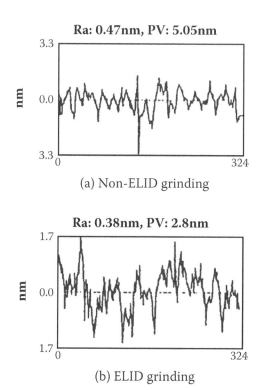

(a) Non-ELID grinding

(b) ELID grinding

FIGURE 9.12 Surface roughness profiles produced by ELID lap grinding and lap grinding (non-ELID) using the #3000000 MRB-D wheel. (a) Non-ELID grinding. (b) ELID grinding.

Figure 9.12 shows the surface roughness profiles produced by ELID lap grinding and lap grinding (non-ELID) using the #3000000 MRB-D wheel. The workpiece used was monocrystalline silicon. The ground surface roughness produced by ELID lap grinding was 2.8 nm in PV and the surface quality was excellent. On the other hand, with lap grinding, the finished surface roughness was 5.05 nm in PV and it was inferior to that by ELID single-side lap grinding.

Figure 9.13 shows an example of the change in removal rate in silicon grinding. Grinding tests were carried out at a wheel speed of 100 rpm and at a work speed of 100 rpm. The applied pressure was 150 kPa. Grinding stabilized after an elapse of 150 minutes without decrease in the removal rate under these operation conditions. Figure 9.14 shows the change in the removal rate of silicon produced by different applied pressures. As expected, the removal rate increased as the applied pressure increased. In either case, grinding was stabilized and mirror-surface finish without any irregularities could be obtained. Owing to the ELID technique, grinding continuity could be achieved with this wheel and ELID lap grinding. Figure 9.15 shows an example of a silicon specimen finished by this system.

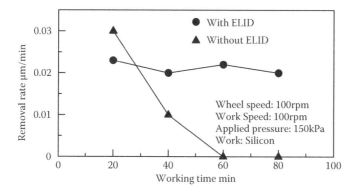

FIGURE 9.13 Effect of ELID on continuity of removal rate when using #3000000 MRB-D wheel.

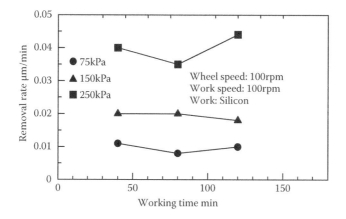

FIGURE 9.14 Effect of applied pressure on continuity of removal rate when using #3000000 MRB-D wheel.

FIGURE 9.15 Example of obtained mirror-surface finish.

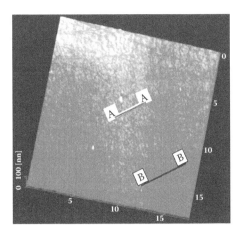

FIGURE 9.16 AFM image by ELID lap grinding using #3000000 MRB-D wheel (A–A: PV1.75 nm; B–B: PV0.82 nm).

Figure 9.16 shows the atomic force microscope (AFM) topographies (area $17 \times 17\ \mu m^2$) of the ground surface of monocrystalline silicon using the #3000000 MRB-D wheel. In the AFM observations, the ground surface produced by this wheel was very smooth with several minute ground grooves crossing one another as shown in Figure 9.16. Though a grain path could be observed on the surface ground, no brittle fracture along the grain path was observed indicating that ductile-mode grinding can be performed by this wheel and ELID single-side lap grinding. In the cross-sectional observation by AFM, the depth of the ground groove was 1.76 nm (A–A) or less (B–B), indicating a very shallow groove. ELID-lap grinding using the #3000000 MRB-D wheel was performed by a small cut not more than several nanometers deep. The surface irregularities of the smooth part were of subnanometer level. The results thus confirmed that surface finishes of subnanometer level can be realized by using this wheel in ELID lap grinding.

9.1.3 ELID Lapping of CVD-SiC

Chemical vapor deposition (CVD)-SiC, which is used for x-ray mirror materials, were ground by ELID single-side lap grinding using #1200–#120000 metal-resin bonded wheels (MRB-D wheels) and the effects of mesh size on the grinding characteristics of this material were studied. After truing the wheels used, grinding characteristics were investigated. First, CVD-SiC was ground using #1200–#1200000 MRB-D wheels, and the effects of grain size on the obtained surface roughness and removal mechanism were investigated. Then, grinding characteristics of CVD-SiC on ELID single-side lap grinding were studied for the purpose of investigating mirror-surface finishing. Grinding tests were carried out at a wheel speed of 100 rpm, workpiece speed of 100 rpm, and applied pressure of 150 kPa.

Figure 9.17 shows the difference in the ground surface roughness by different mesh numbers. The surface finish improved as the grain diameter decreased. With this method and use of the #120000 MRB-D wheel, the surface could be

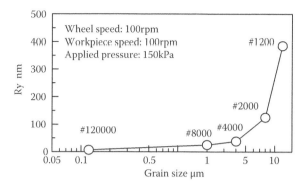

FIGURE 9.17 Relation between grain size and surface roughness.

ground to an excellent finish of 6.5 nm in Ry. Figure 9.18 shows the scanning electron microscope (SEM) photographs of the ground surface of CVD-SiC. The surface ground by the #1200 MRB-D wheel shows typical brittle fracture removal in the SEM observation. On the surface ground by the #2000 MRB-D wheel, a small brittle fracture could be observed on the surface, but this surface was more or less smooth. The surfaces ground by the #4000, #8000, and #120000 MRB-D wheels are smoother than that by the #2000 wheel and no brittle fracture was seen. With this material, transition from the brittle to ductile mode was achieved using wheels over #4000 with ELID lap grinding.

The surface ground by the #120000 MRB-D wheels was observed by AFM. Figure 9.19 shows the AFM images of the ground surfaces, which indicate that the ground surface is very smooth and consists of fine grinding marks that cross one another. In the analysis of the cross-section of parts A–A and B–B in Figure 9.19, the depth of the grinding mark (A–A) is found to be extremely shallow (approximately 4.3 nm) and the surface irregularities of the smooth part (B–B) were 2.9 nm. The experiment confirmed that ELID lap grinding using this wheel enables mechanical removal of materials in the order of several nanometers. Figure 9.20 shows the mirror surface finishing.

For the purpose of grasping the performance of ELID single-side lap grinding using #4000, #8000, and #120000 MRB-D wheels, CVD-SiC was ground by this system, and the grinding efficiency and surface roughness were investigated. Figure 9.21 shows the stability of the grinding performance. For all wheels, the stock removal increased linearly as the grinding time increased, indicating that stable grinding was achieved without clogging. Like the results of stock removal, the surface roughness values were stable during grinding tests. These results show that ELID lap grinding can realize the stable use of fine grit diamond wheels and is a very useful method for stable mirror-surface finish.

The relation between the grinding time and surface roughness was studied. In this experiment, rough grinding was performed using the #1200 MRB-D wheel, followed by mirror-surface finishing using #4000, #8000, and #120000 MRB-D wheels, and the

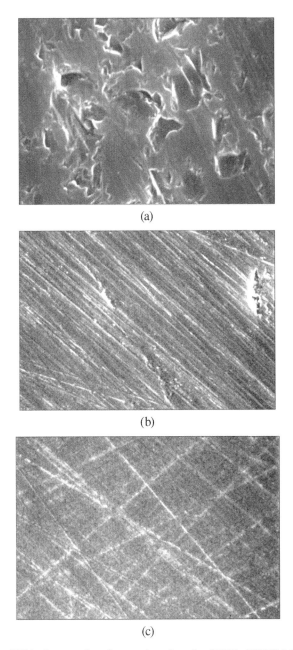

FIGURE 9.18 SEM photographs of ground surface by #1200–#120000 MRB-D wheel.
(a) #1200 MRB-D wheel. (b) #2000 MRB-D wheel. (c) #4000 MRB-D wheel. (d) #8000
MRB-D wheel. (e) #120000 MRB-D wheel.

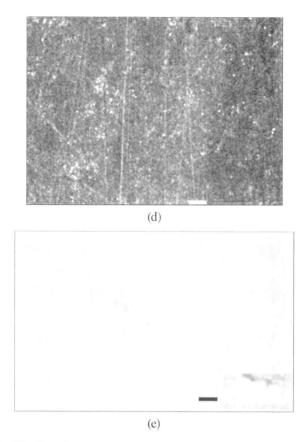

(d)

(e)

FIGURE 9.18 (Continued).

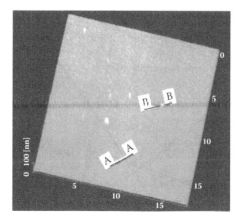

FIGURE 9.19 AFM image of CVD-SiC ground by #1200000 MRD-D wheel.

FIGURE 9.20 Example of obtained mirror-surface finish.

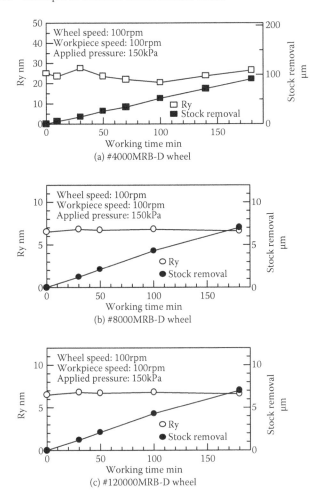

FIGURE 9.21 Stability of grinding performance on ELID single-side lap grinding. (a) #4000 MRB-D wheel. (b) #8000 MRB-D wheel. (c) #120000 MRB-D wheel.

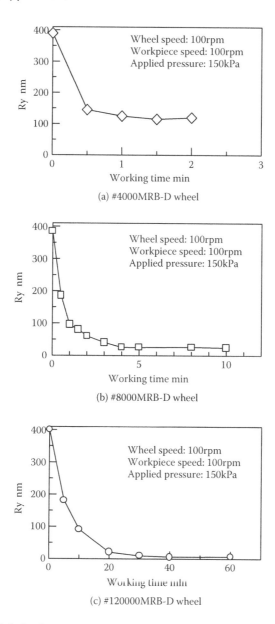

FIGURE 9.22   Relation between working time and surface roughness. (a) #4000 MRB-D wheel. (b) #8000 MRB-D wheel. (c) #120000 MRB-D wheel.

change in the obtained surface roughness was investigated. Figure 9.22 shows the relation between the grinding time and obtained surface roughness. The workpieces used were ground by the #1200 MRB-D wheel to equalize the initial conditions before testing. The obtained surface roughnesses were about 400 nm Ry. In the case of the #4000 MRB-D wheel, when grinding starts, the surface roughness rapidly improved within

1 minute of grinding, and then the surface roughness showed constant values. For the #8000 MRB-D wheel, like the results of the #4000 MRB-D wheel, the surface roughness rapidly improved within 2 minutes of grinding, and then the surface roughness gradually improved and stabilized. After 5 minutes of grinding, the surface roughness improved from 395 nm Ry to 25 nm Ry. The results indicate that efficient mirror-surface grinding can be realized in this system. On the other hand, for the #120000 MRB-D wheel, although the surface roughness improved when grinding started, it took a long time for the surface roughness to show constant values. In this case, efficient mirror-surface finishing could not be achieved under these conditions, but it must be realized to perform middle grinding using the #4000 or #8000 MRB-D wheel after rough grinding.

To realize efficient grinding of mirror and lens production, grinding characteristics of CVD-SiC in ELID single-side lap grinding using different grain size wheels were studied. Through SEM observation, it was clarified that transition from the brittle to ductile mode was achieved using wheels over #4000 MRB-D with ELID lap grinding. The experimental results using the #4000, #8000, and #120000 MRB-D wheels indicate that stable and efficient mirror-surface finishing can be achieved in this system. The AFM image showed that ELID lap grinding using this wheel successfully removes materials mechanically in the order of several nanometers.

9.1.4 Efficient and Smooth Grinding Characteristics of Monocrystalline 4H-SiC Wafer

Nowadays most semiconductors are monocrystalline silicon (Si) and are used in a broad range of applications such as electrical appliances and cellular phones. In particular, the semiconductors used for power conversion are generically called power devices. Power devices also have a broad range of uses, such as in power plants, electric trains, hybrid and electric cars, and air conditioners. Therefore, improving power devices contributes to saving electric power and reducing carbon dioxide (CO_2). Although it has improved, the performance of power devices made of Si will be limited in the near future.[4]

On the other hand, silicon carbide (SiC) materials have increasingly been used in a wide range of industries because of excellent mechanical and electrical properties. Currently, SiC power devices have been gaining attention as next-generation semiconductors because they have excellent electrical characteristics. Specifically, SiC power devices surpass Si power devices in power conversion efficiency, switching speed, and high-temperature operation.[5] However, SiC has difficulties in efficient and smooth grinding because of its hard and brittle character. For instance, Vickers hardness (Hv) of Si is 1000, whereas SiC has a Vickers hardness of 2400 to 2700.[6] Therefore, it is necessary to grind SiC efficiently for diffusion of SiC power devices. Monocrystalline SiC has some polytypes, which are usually expressed by a number and a letter. The number shows atomic stacking sequences along the c axis and the letter shows a crystal system. For instance, "4H" stands for atoms crystallized in a hexagonal system and the unit consists of four atoms. At present, 4H-SiC and 6H-SiC are commonly used for SiC power devices because they have less anisotropy and higher electron mobility than other types of SiC.[7] Figure 9.23 shows crystal structures of 4H-SiC and 6H-SiC. A white circle represents an Si atom and a black circle represents a C atom. A letter represents an

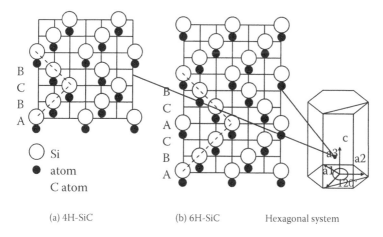

(a) 4H-SiC   (b) 6H-SiC   Hexagonal system

FIGURE 9.23 Crystal structures of 4H-SiC and 6H-SiC. (a) 4H-SiC. (b) 6H-SiC Hexagonal system.

atomic stacking structure along the c axis. In Japan, SiC power devices are commonly made of 4H-SiC because it has less anisotropy, higher electron mobility, and higher breakdown voltage than 6H-SiC.[8] Figure 9.24 shows a typical crystallographic plane of 4H-SiC. The (0001) surface was produced relatively more than the (11 −20) surface. So, the workpiece we used was a 2-inch 4H-SiC wafer, which is a domestic product, and the crystallographic plane was (0001). Figure 9.25 shows an external view of that 2-inch 4H-SiC wafer, which is 51.1 mm in diameter and 0.5 mm thick.

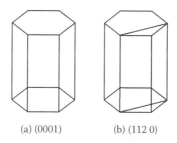

(a) (0001)   (b) (112 0)

FIGURE 9.24 Typical crystallographic planes of 4H-SiC. (a) (0001). (b) (1120).

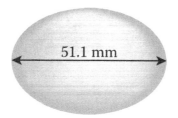

51.1 mm

FIGURE 9.25 External view of 4H-SiC wafer.

Before investigating grinding conditions, we determined surface roughness of the workpiece. Figure 9.26 shows surface roughness of the 4H-SiC wafer measured by NewView (Zygo Corp.) optical profiler. The root-mean-square height (Rq) was 578 nm, and the peak-to-valley distance (PV) was 4370 nm. Considering the results of the previous work[9,10] and the balance between removal volume and surface roughness, we started machining with the #1200 mesh wheel and set 40 μm for the amount of grinding. Table 9.1 shows the principal experimental conditions. With regard to wheel rotation and workpiece rotation, we investigated some efficient combinations of high and low circumferential velocities. To clarify the effect of wheel rotation and workpiece rotation, the feed rate was fixed at 1 μm/min. After each machining, we determined a removal volume and surface roughness. The removal volume was evaluated by measuring removal height with a dial gauge. The surface roughness was evaluated by the same NewView profiler. The measure point was the middle of the center and the edge of the wafer.

An appropriate condition we investigated with the #1200 mesh wheel was as follows:

- Wheel rotation—1000 min⁻¹ (circumferential velocity: 449 m/min)
- Workpiece rotation—300 min⁻¹
- Feed rate—1.0 μm/min
- Open voltage—60 V
- Peak current—4 A

We machined three times under these conditions and determined the average removal volumes, which was 32.0 μm when the amount of grinding was 40 μm; this indicates 80.0% in the amount of grinding. Surface roughness was 7.74 nm Rq and 74.0 nm PV. Figure 9.27a shows the surface roughness using the NewView profiler. In addition, we measured 2.54 nm Rq and 65.9 nm PV near the wafer edge. Those results indicate that the 4H-SiC wafer was ground precisely and efficiently with the #1200 mesh wheel.

Concerning the removal volume, fast workpiece rotation had a tendency to efficiently grind the workpiece. Low open voltage and low peak current also tended to efficiently grind the workpiece. However, wheel rotation did not have much influence on removal volume in this investigation. With regard to surface roughness, slow workpiece rotation had a tendency to grind the workpiece smoothly. Low open voltage and high peak current also tended to grind the workpiece smoothly. Moreover, fast wheel rotation had slight tendency to grind the workpiece smoothly in this investigation.

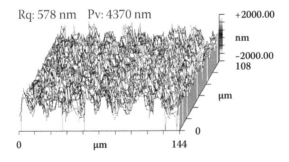

FIGURE 9.26   Surface roughness before grinding.

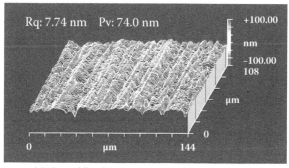

(a) The middle of the center and the edge of the wafer

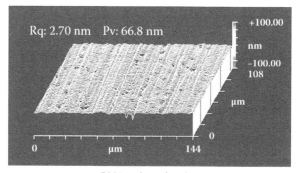

(b) Near the wafer edge

FIGURE 9.27 Surface roughness after #1200 grinding. (a) The middle of the center and the edge of the wafer. (b) Near the wafer edge.

When the other conditions are constant, much greater removal volume requires fast workpiece rotation, but fast workpiece rotation makes surface roughness coarse. On the other hand, smooth surface roughness needs low open voltage and high peak current, but low open voltage and high peak current have little influence on removal volume.

Figure 9.28 shows an external view of the workpiece after #1200 grinding. Though there were some grinding traces on the workpiece, the surface is becoming a mirror-like surface. Figure 9.29 shows an SEM [JSM-5600LV; JEOL Ltd.] image of the workpiece ground with the #1200 mesh wheel. There were some air spaces and grinding traces

FIGURE 9.28 An external view of the workpiece after #1200 grinding.

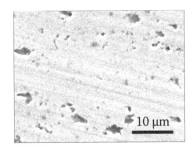

FIGURE 9.29 Observation by SEM after #1200 grinding (×1500).

in a ductile mode on the surface. Considering the size of air spaces and the property of the surface, finer mesh wheels will decrease the size of air spaces and improve the surface.

9.2 ELID POLISHING AND GRINDING OF MAGNETIC HEADS

9.2.1 INTRODUCTION

ELID polishing is a method of constant pressure grinding that employs the electrically conductive bonded wheel and the ELID.[11] The ELID technique is a novel technique that was pioneered by Dr. Hitoshi Ohmori of the Institute of Physical and Chemical Research in Tokyo, Japan. With the application of this technique the metal bonded wheel is electrolytically dressed during the process.[15] ELID is the method using electrolysis, where the worn abrasive grains are removed and new grains are caused to protrude from the wheel surface.

9.2.2 EXPERIMENTAL SETUP

The experiments for ELID polishing of magnetic heads were carried out on a Warren lapping machine. This is a traditional lapping machine, which was modified for ELID using suitable fixtures.

The ELID setup is shown in Figure 9.30 and it consisted of the following:

- Electrode
- Polishing wheel

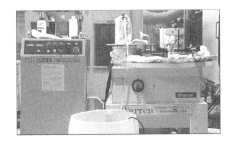

FIGURE 9.30 Experimental setup for ELID polishing.

FIGURE 9.31   Electrode used for ELID polishing.

- Electrically conductive brush
- ELID power supply
- ELID fluids

The electrode is made of copper (110 copper with 99% purity). This electrode is attached to the negative terminal of the ELID power supply in the electrolysis. The electrode is shown in Figure 9.31. Copper is chosen as the electrode because of its excellent electrical conductivity (which can be matched only by silver), reasonable cost, availability in the market, and good machinability. While designing the shape and size of the copper electrode the following requirements were to be followed for ELID.

As per the requirement for ELID the electrode has to cover one-sixth of the peripheral length of the polishing wheel. Hence, the electrode is designed with an included angle of 60 degrees.

The position of the electrode is another critical parameter that has to be followed for obtaining good results in ELID. The electrode has to be held above the rotating polishing wheel at a distance of 0.1 mm to 0.3 mm. In order to achieve this, a suitable fixture was constructed. This fixture is shown in Figure 9.32. The fixture consists of the following items:

Supporting plate—This mild steel plate is mounted on a cylinder located on the polishing machine. The plate is rigid enough to hold the weight of the copper electrode and the length is such that it can position the copper electrode accurately in the horizontal plane above the rotating polishing plate.

Supporting rollers and springs—Cylindrical nylon pieces are used to hold the copper electrode to the supporting plate. Nylon is used because it has a good machinability, strength, and above all it is a nonconducting material. Hence, it can prevent the flow of the current from the copper electrode to the machine. Springs are inserted between the supporting plate and the roller. They help in damping the vibrations generated during the process and also in preventing the horizontal movement of the rollers. Fine threaded screws are used to hold the rollers to obtain precise positioning of the electrode.

Arrangement for supplying ELID fluid—For achieving good ELID there has to be a continuous film of ELID fluid between the electrode and the polishing wheel. This is achieved by supplying the fluid through pipes, which are attached on the electrode as shown in Figure 9.32. The pipes are arranged in such a way that the fluid is evenly distributed.

An acrylic insulator is used to prevent the current from passing to the machine through the electrically conductive polishing plate. The insulator is fitted between the drive shaft and the polishing wheel. The insulator is shown in Figure 9.33.

A polishing wheel made of tin (CI base and tin top) is used for the experiments. The plate is 18 inches in diameter and 4 inches thick. The polishing wheel after the experiments is shown in Figure 9.34.

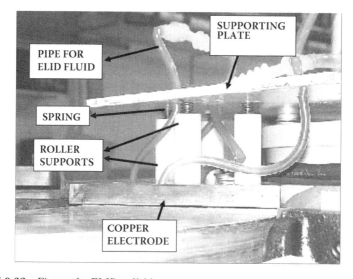

FIGURE 9.32 Fixture for ELID polishing.

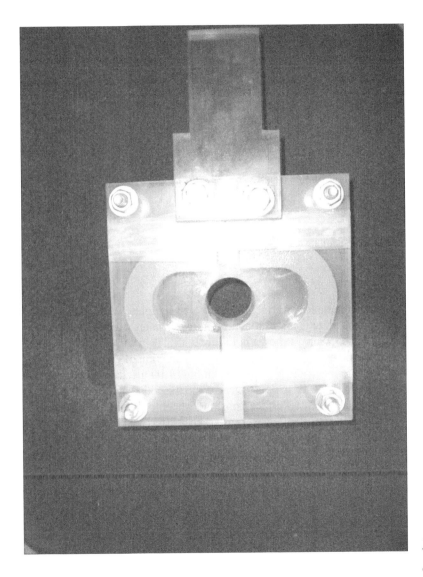

FIGURE 9.33 Insulator.

FIGURE 9.34   The polishing wheel.

A fixture had to be made for attaching the polishing plate to the drive shaft, as the diameter of the plate is larger than the standard plates used for the Warren lapping machine. The fixture is shown in Figure 9.35. The fixture lifts the polishing plate above the machine bed and avoids the conduction of electricity.

The polishing plate forms the positive pole in the ELID mechanism. An electrically conductive brush was used to pass the current to the polishing wheel. The brush has to be in continuous contact with the polishing wheel for the ELID mechanism. This is achieved by using a spring-loaded brush. The brush is held in a vice to prevent vertical or horizontal movement due to contact with the rotating polishing wheel.

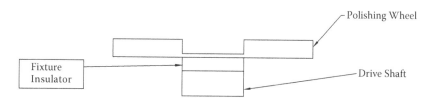

FIGURE 9.35   Schematic showing position of the fixture.

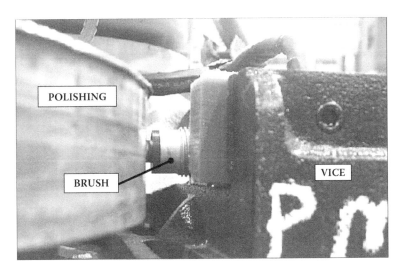

FIGURE 9.36 Setup for the electrically conductive brush.

The vice is clamped to the bed with the help of two C-clamps. The setup is shown in Figure 9.36.

ELID polishing requires a special power supply, which was obtained by using the Fuji ELIDer. The Fuji ELIDer is capable of generating a current up to 90 A and voltages up to 130 V. The Fuji ELIDer is shown in Figure 9.37. The experimental conditions are listed in Table 9.3.

TABLE 9.3
Experimental Conditions

Grinding machine	Rotary in-feed grinding machine	HGS-10A2
Grinding wheels	Cast-iron bonded diamond wheel	SD#1200
		Concentration 100
		(φ143 × W3 mm cup)
Grinding conditions	Wheel rotation	1000, 2500 min^{-1}
		(Circumferential velocity:
		449, 1120 m/min)
	Workpiece rotation	100, 300 min^{-1}
	Feed rate	1.0 µm/min
	Amount of grinding	40 µm
Power supply		NX-ED911
Electrical conditions	Open voltage	60, 90 V
	Peak current	4, 10 A
	Pulse timing (on/off)	2/2 µs

FIGURE 9.37   Fuji ELIDer.

For ELID, three types of fluids were used:

Diamond abrasive fluid—This was used for dressing the wheel with the dia-
    mond abrasives. It is a 0.25 ct diamond abrasive fluid.
ELID fluid—This is a special type of fluid prepared especially for ELID.
    It is a high conductivity fluid, which helps in better electrolysis and also
    performs the other functions of a typical polishing fluid. The fluid forms

TABLE 9.4

Surface Roughness of Plate after Facing

	Ra	R	Rq	Rv	Rmax	Rq	Rz	Ry
Average uin	56.856	182.182	65.238	154.8	262.388	65.238	250.27	277.622
Average um	1.444	4.627	1.657	3.932	6.665	1.657	6.357	7.052

Note: Ra, average roughness; Rq, root mean square roughness; Rv, maximum profile valley depth; Rmax, maximum roughness depth; Rz, average maximum height of the profile; Ry, maximum height of the profile.

a thin layer between the electrode and the polishing wheel and brings about electrolysis.

Lubricant—While conducting the experiments for ELID polishing it was observed that the plate gets damaged due to the high current and voltage and hence an olive oil was used as a lubricant. The lubricant is fed at the rate of 2 milliliter every 15 seconds.

Before conducting the experiments for ELID polishing the following steps were taken to obtain accurate results. Facing the polishing wheel is done to bring surface roughness of the polishing wheel to a desirable level. The surface roughness of the wheel was checked with the Hommel profilometer. To obtain a good and even surface the plate was faced on a lathe. The surface was rechecked and the results were shown in Table 9.4.

The diamond abrasives (0.25 monocrystalline diamonds) are embedded into the polishing wheel before ELID polishing. Charging is an elaborate procedure, which has to be done for 120 minutes. The process is shown in Figure 9.38. The abrasive slurry containing diamond abrasives is fed to the rotating polishing plate at the rate of 0.16 ml/sec into one of the conditioning rings. The other conditioning rotates with

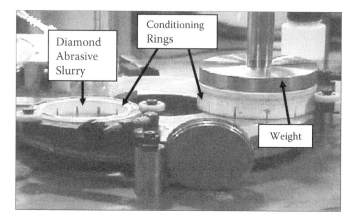

FIGURE 9.38 Charging of the diamond abrasives.

FIGURE 9.39 Mounting the parts.

a weight of 12 kg in it. As the conditioning ring rotates the diamond abrasives get embedded into the soft tin plate when they come below the conditioning ring and the weight. The rotating conditioning ring also ensures even distribution of the grains over the surface of the plate. After charging the grains the actual process is started. The gap between the copper electrode and the polishing wheel is checked with a slip gauge. The parts are mounted as shown in Figure 9.39. Six parts are mounted each time. Before mounting the parts the parts are cleaned with alcohol.

After this the parts are mounted in the conditioning rings and the parameters, like the current and voltage, are set as desired. The polishing machine is started and the ELIDer is switched on. The machine is set for a time limit of 6 minutes, which is the time for each experiment.

After the experiment is over the parts are carefully removed from the machine and washed in alcohol to remove any contaminants and unwanted material that gets deposited during the polishing process. The thickness of the parts is again measured with a micrometer and tabulated.

9.2.3   ELID PARAMETERS

Some of the important parameters that affect the surface finish or the material removal rate obtained by the ELID process are as follows:

9.2.3.1   Grain Size or Mesh Number of the Wheel

The grinding wheel forms one of the most important parts in ELID lap grinding and the characteristics of the wheel considerably affect the surface finish and material removal rate. The grinding wheels can be resin bonded, vitrified bonded, or metal bonded. Figures 9.40 and 9.41 show the effect of mesh number on surface finish. For both materials the surface roughness and material removal rate decreased.

As the mesh number increases the material removed by the grain decreases and the surface finish improves. Average roughness does not improve once the depth of cut is decreased to a minimum level for a particular grain size. Therefore, with consideration of minimum available depth of cut, appropriate grinding wheels with proper mesh size should be selected for a grinding process.[14]

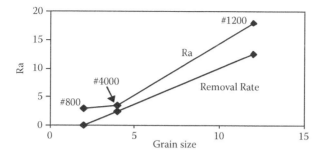

FIGURE 9.40 Effect of grain size on Ra and removal rate for tungsten carbide. (From N. Itoh, H. Ohmori, and B. P. Bandopadhyay, *International Journal for Manufacturing Science and Production* 1, no. 1 (1997), 9–15. With permission.)

9.2.3.2   Truing Time

Truing is an integral part of ELID and has to be done to reduce the eccentricity of the wheel. Metal-bonded diamond grinding wheels are efficiently trued using a new method known as electrodischarge (ED) truing. The relationship between truing time and material removal rate is shown by Ohmori in Figure 9.42.[11]

9.2.3.3   Cutting Speed

In a series of experiments conducted by E. Lee effects of cutting speed on surface finish were studied. The experiment is conducted at a workpiece feed rate of 80 mm/min and a depth of 1 µm per pass. The results are shown in Figure 9.43. From the figure it can be said that cutting speed has no significant influence on the surface finish of the workpiece.[16]

9.2.3.4   Grinding Force

Grinding force decreases when using optimum in-process electrolytic dressing for grinding the workpiece. In conventional grinding as material removal rate (MRR) increases, the normal grinding force also increased proportionally. This is shown in Figure 9.44.[15]

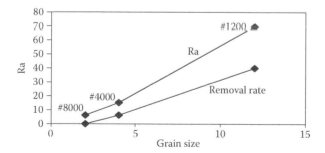

FIGURE 9.41 Effect of grain size on Ra and removal rate for silicon. (From N. Itoh, H. Ohmori, and B. P. Bandopadhyay, *International Journal for Manufacturing Science and Production* 1, no. 1 (1997), 9–15. With permission.)

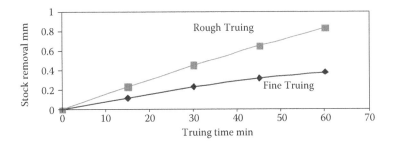

FIGURE 9.42 Effect of truing time on surface roughness. (From N. Itoh, H. Ohmori, and T. Karaki-Doy, *Journal of Materials Processing Technology* 62 (1996), 315–320. With permission.)

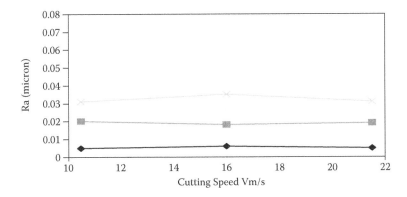

FIGURE 9.43 Effect of cutting speed on Ra. (From E.-S. Lee, *Journal of Materials Processing Technology*, 100 (2000), 200–208. With permission.)

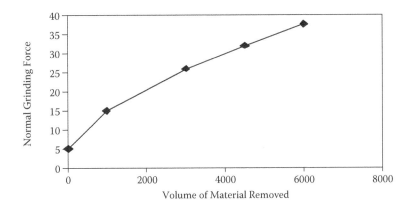

FIGURE 9.44 Variation of grinding force with MRR in conventional grinding. (From B. P. Bandopadhyay, *Abrasives* (1997, April/May), 10–34. With permission.)

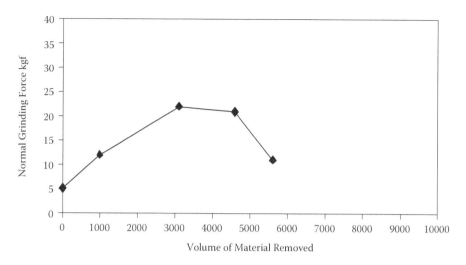

FIGURE 9.45 Variation of grinding force with MRR in ELID. (From B. P. Bandopadhyay, *Abrasives* (1997, April/May), 10–34. With permission.)

The variation of grinding force with MRR in ELID grinding is shown in Figure 9.45. When we look at both curves, continuous increase in grinding force is observed in conventional grinding. The grinding force reached around 38 kgf (373 N) when the volume of material removed reached 6000 mmñ. While in ELID, as the volume of material increased, the grinding force decreases. This effect is more visible after 4500 mmñ. The effects of ELID grinding were studied by performing ELID grinding with voltage increased to 90 and I_p to 24 A. The results as obtained by B. P. Bandopadhyay were presented in Figure 9.46.

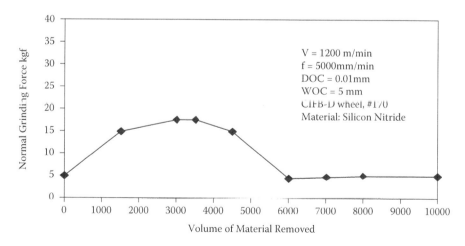

FIGURE 9.46 Variation of grinding force with MRR in ELID. (From B. P. Bandopadhyay, *Abrasives* (1997, April/May), 10–34. With permission.)

TABLE 9.5
Electrical Resistivity of Wheel Developed

Metal:Resin	Electrical Resistivity (ohm-mm)
3:7	1733.33
5:5	0.32
7:3	0.13

Source: N. Itoh, H. Ohmori, T. Kasai, and T. Karaki-Doy, *International Journal of Electric Machining* 3 (1998), 13–18. With permission.

Grinding force stabilizes after 6000 mm ñ of material removal and almost remained constant. Thus, it can be concluded that the full potential of ELID grinding, that is, reduced grinding force, can be utilized only after the process has been stabilized.

9.2.3.5 Wheel Resistivity

The electrical resistivity has a tremendous effect on the characteristics of ELID. Experience has shown that the wheel can be used for ELID if the resistivity is below 0.5 ohm/mm. Table 9.5 shows the electrical resistivity of the wheel developed with different combinations of metal and resin. It is seen that as the percentage of resin in the mixture increases, the electrical resistivity also increases. The results of a series of experiments have shown that a mixing ratio of 7:3 produces good electrical conductivity and stable grinding.[14]

9.2.3.6 Wheel Material

Wheel material also affects the performance of ELID. Typical results are shown in Figure 9.47. The removal rate of the metal-resin-bonded wheel was inferior to that of the cast-iron bonded wheel, but the former produced better surface roughness. The reason for

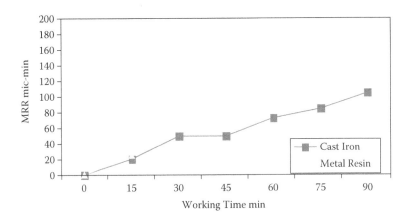

FIGURE 9.47 Variation of MRR in MRB and CIFB wheels. (From N. Itoh, H. Ohmori, T. Kasai, and T. Karaki-Doy, *International Journal of Electric Machining* 3 (1998), 13–18. With permission.)

this is that as elastic deformation is likely to occur with metal and resin bonded wheels than the cast-iron wheel, evenness and reduction in depth of cut were achieved.[12]

9.2.4 Results and Discussions

9.2.4.1 ELID Polishing Results

After testing the feasibility of ELID polishing, experiments were conducted to obtain more information and to compare it with the normal polishing process. The parameters used for ELID polishing were

Variables—Current, voltage, and load
Time—6 min
Lubricant feed rate—0.01 ml/sec
Workpiece material—AlTiC magnetic head
Load used—3 kg and 5 kg
Abrasives used—0.25 micron monocrystalline diamond

For the analysis the parts were inspected under an AFM and the surface roughness parameters such as Ra and RMS were calculated. Also, the relation between material removal and parameters like current, voltage, and load was observed by plotting graphs.

The variation of material removal with current can be summarized in Figure 9.48. The parameters were

Voltage—60 V
Current—30 A and 40 A
Workpiece—AlTiC magnetic head
Load used—3 kg and 5 kg
Abrasive used—0.25 micron monocrystalline diamond
Time—6 minutes

The material removal increases with an increase in current. As the current increases, the dressing speed of the wheel also increases, and consequently fresh grains are exposed more rapidly. This causes more material removal.

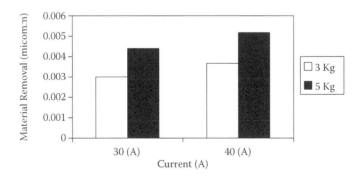

FIGURE 9.48 Material removal and current.

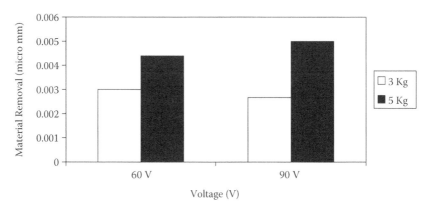

FIGURE 9.49   Material removal and voltage.

The relation between material removal and voltage is shown in Figure 9.49.

Voltage—60 V and 90 V
Current—30 A
Workpiece—AlTiC magnetic head
Load used—3 kg and 5 kg
Abrasive used—0.25 micron monocrystalline diamond
Time—6 minutes

The material removal increases with the increase in the voltage. As the voltage increases, the dressing speed of the wheel also increases, and consequently fresh grains are exposed more rapidly. This causes an increase in material removal.

The relation between material removal and voltage is shown in Figure 9.50.

Voltage—60 V
Current—40 A
Workpiece—AlTiC magnetic head
Load used—3 kg and 5 kg
Abrasive used—0.25 micron monocrystalline diamond
Time—6 minutes

The material removal is more with ELID as expected, which proves the feasibility and success of ELID Polishing. Due to in-process dressing fresh abrasive grains are continuously exposed and consequently material removal is higher. It also shows that in ELID a lower force produces more material removal.

Figure 9.51 illustrates how AFM works. As the cantilever flexes, the light from the laser is reflected onto the split photodiode. By measuring the difference signal (A–B), changes in the bending of the cantilever can be measured. Since the cantilever obeys Hooke's law for small displacements, the interaction force between the tip and the sample can be found. The movement of the tip or sample is performed by an extremely precise positioning device made from piezoelectric ceramics, most often in the form of a tube scanner. The scanner is capable of subangstrom resolution in x,

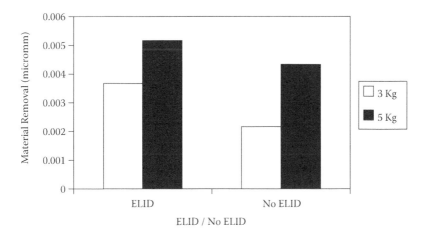

FIGURE 9.50 Material Removal and ELID.

y, and z directions. The z axis is conventionally perpendicular to the sample. AFM operates by measuring attractive or repulsive forces between a tip and the sample. In its repulsive "contact" mode, the instrument lightly touches a tip at the end of a leaf spring or "cantilever" to the sample. As a raster scan drags the tip over the sample, some sort of detection apparatus measures the vertical deflection of the cantilever, which indicates the local sample height. Thus, in contact mode the AFM measures hard-sphere repulsion forces between the tip and sample.

The surface roughness parameters were measured using an AFM. The results obtained were plotted on Figure 9.52. The parameters were

Voltage—60 V
Current—30 A and 40 A
Workpiece—AlTiC magnetic head
Load used—3 kg and 5 kg

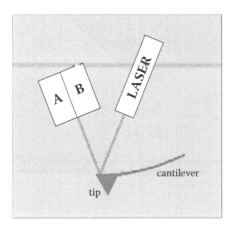

FIGURE 9.51 AFM principle.

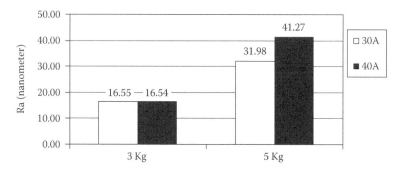

FIGURE 9.52 Surface roughness (Ra) and current.

Abrasive used—0.25 micron monocrystalline diamond
Time—6 minutes

The surface deteriorates as the load and the current increases. This can be attributed to the fact that the current was set to a very high value and hence it caused higher surface roughness. Surface profile and surface roughness obtained by varying the current are shown in Figures 9.53 and 9.54, respectively.

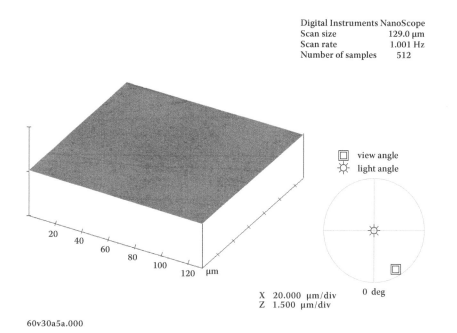

FIGURE 9.53 Surface profile obtained by varying current.

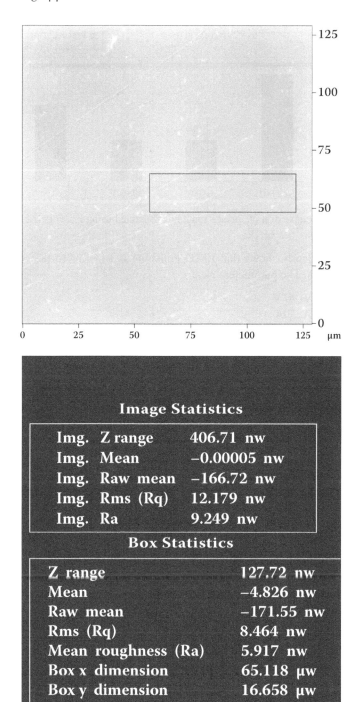

FIGURE 9.54 Surface roughness obtained by varying current.

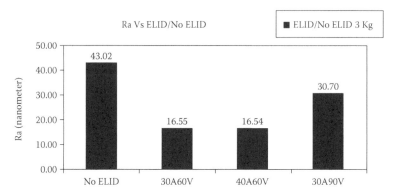

FIGURE 9.55 Comparison of surface roughness obtained using ELID polishing and conventional polishing.

Figure 9.55 clearly shows that ELID polishing is superior to the conventional polishing process. The parameters were

Voltage—60 V, 90 V
Current—40 A, 30A
Workpiece—AlTiC magnetic head
Load used—3 kg and 5 kg
Abrasive used—0.25 micron monocrystalline diamond
Time—6 minutes

9.2.4.2 Statistical Analysis of the Results

Statistical tests such as the t-test and the F-test were used to analyze the data obtained in ELID polishing. These tests are used to formulate inferences about population parameters and to validate theory against observation. For the tests conducted, the data collected was assumed to have normal distribution. The t-test and F-test are used for testing of mean and variability, respectively. For the t-test the t-statistic is given by

$$t = \frac{(\bar{x}_1 - x_2)}{S_p\sqrt{\frac{1}{n_1} + \frac{1}{n_2}}}$$

where,

$t = t$ statistic
x_1, x_2 = Mean of sample 1 and sample 2, respectively
S_p = Standard deviation
n_1, n_2 = Sample size of the two samples

For testing H_0: $u_1 - u_2 = 0$ and H_a: $u_1 - u_2 = 0$ and the degree of freedom is $n_1 + n_2 - 2$. If the t obtained from the formula is greater than t from the distribution table, then the hypothesis is rejected.

For the F-test the test statistic is given by

$$F = \frac{S_1^2}{S_{21}^2}$$

TABLE 9.6
F-Test for Change in Material Removal with Current

	Variable 1	Variable 2
Mean	0.003	0.0044
Variance	6.66667E–07	2.3E–06
Observations	4	5
Degree of freedom	3	4
F	0.289855072	
P(F <= f) one-tail	0.168437949	
F critical one-tail	0.109682929	

where,

F = Test statistic

S_1, S_2 = Sample variances

In the F-test, H_0: $\sigma_1^2 \leq \sigma_2^2$ versus H_a: $\sigma_1^2 > \sigma_2^2$. The hypothesis is accepted if $F < F$ (a, b), where a and b are degrees of freedom.

For the F-test, the hypothesis is formulated as follows:

σ_1 = Standard deviation of the material removal for 3 kg load

σ_2 = Standard deviation of the material removal for 5 kg load

H_0: $\sigma_1^2 \leq \sigma_2^2$ and H_a: $\sigma_1^2 > \sigma_2^2$, $\alpha = 0.05$

Table 9.6 shows the F-test for change in material removal with the current. The results show that $F < F$ (a, b), hence the sample with the 5 kg load has less variability.

For the t-test the hypothesis is formulated as follows:

μ_1 = Mean for the material removal for 3 kg load

μ_2 = Mean for the material removal for 5 kg load

H_0: $\mu_1 - \mu_2 = 0$ and H_a: $\mu_1 - \mu_2 \# 0$, $\alpha = 0.05$

From Table 9.7, we see that the hypothesis is accepted and hence we can conclude that the results obtained from the 3 kg load are better than that of 5 kg.

TABLE 9.7
t-Test for Change in Material Removal with Current

	Variable 1	Variable 2
Mean	0.003	0.0044
Variance	6.66667E–07	2.3E–06
Observations	4	5
Hypothesized mean difference	0	
df	6	
t Stat	–1.768519034	
t Critical one-tail	1.943180905	

TABLE 9.8
F- and t-Test for Change in Material Removal

	F-Test	t-Test
Voltage	Rejected	Accepted
Load	Rejected	Accepted
Polishing	Rejected	Accepted

Similar tests were conducted for the other sets of readings and the results are shown in Table 9.8.

F-tests for all the samples fail, which indicates that the null hypothesis is untrue. The variability of the material removal rate by the 3 kg load is more than that by the 5 kg load. The t-tests are accepted for all the tests, which indicates that the hypothesis is true proving that the means of the material removal rate due to 3 and 5 kg are the same.

9.3   MOLDS

Experiments to generate a lens mold profile for tungsten carbide (WC) and SiC materials were attempted by using a four-axis ultraprecision grinding machine. Workpieces were mounted on the C axis of the ultraprecision grinding machine through a vacuum chuck. This ultraprecision grinding machine also has an ELID grinding function.

Prior to lens mold profile generation, a truing operation was conducted for the grinding wheels. This truing operation is shown in Figure 9.56 and was electrically done between a rotary truer and a grinding wheel. This method is called plasma discharge truing and is an effective truing method for iron-bonded grinding wheels especially used for the ELID grinding method. The positive pole was applied to the grinding wheel, the negative pole was applied to the rotary electrode, and a pressurized mist was supplied in between. Better efficiency and accuracy can be obtained by this plasma discharge truing method. Figure 9.57 shows the microscope observation of the grinding wheel surfaces after the plasma discharge truing. Uniform truing was done on the center area of the grinding wheel width, which is especially used for lens mold profile generation. As workpieces, WC and SiC were used, and four kinds of

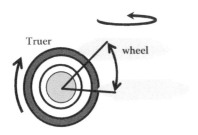

FIGURE 9.56   Plasma discharge truing of wheel.

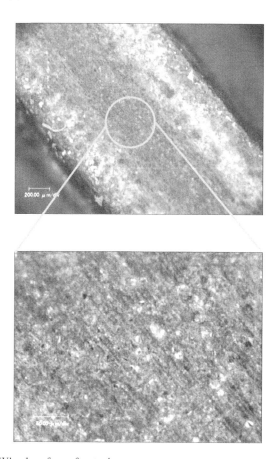

FIGURE 9.57 Wheel surface after truing.

diamond grinding wheels from #325 to #8000 were applied. Specific software was applied to generate a tool path for lens mold profiles.

Figure 9.58 shows a schematic illustration and a close-up view of the setup for lens mold grinding. Cross-grinding was applied, in which the grinding wheel scans in the direction of the diameter of lens mold. Experiments were conducted with reduced ELID conditions in order to reduce grinding wheel wear and to keep the

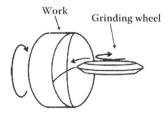

FIGURE 9.58 Microlens mold grinding.

TABLE 9.9
Roughness

Wheel	PV (nm)	Ra (nm)
#1200	848.7	17.7
#325	1894.4	109.3

grinding wheel profile. Roughness by SD#325 and SD#1200 are shown in Table 9.9, and a profile by SD#1200 is shown in Figure 9.59.

Roughness patterns are shown in Figure 9.60. SD#1200 produced a significantly improved surface compared with SD#325. A profile after compensation for the NC program for tool path is shown in Figure 9.61. Roughness by SD#4000 is added in Table 9.10. More improvement can be confirmed.

Difference in surface quality is shown in Figure 9.62. The difference seems to be attributed to a difference in peripheral speed on grinding. Roughness by SD#8000 is added in Table 9.11. View of ground mold sample on WC is summarized in Figure 9.63. Roughness pattern by SD#8000 is shown in Figure 9.64, and a view of ground mold samples on SiC are shown in Figure 9.65.

9.4 MICROTOOLS

9.4.1 INTRODUCTION

Microtools having an outer diameter, at the tip, of less than several tens of microns play a leading role in "cutting edge" research and in the development of new industrial technologies. For example, microtools support a wide variety of needs, such as the microfabrication of various tools used for semiconductor devices, microlens arrays, measurement microprobes, and biomanipulators. Meanwhile, the fabrication of microtools presents another challenge in micromachining. Imprecise geometry and the irregularity of tools often negate the advantages of ultrafine

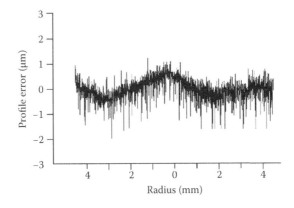

FIGURE 9.59   Profile by SD#1200 without compensation.

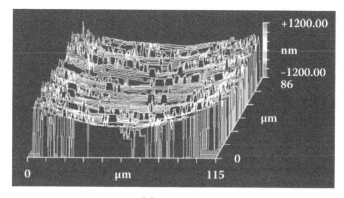

(a) SD#325

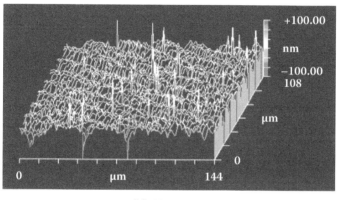

(b) SD#1200

FIGURE 9.60   Roughness patterns. (a) SD#325. (b) SD#1200.

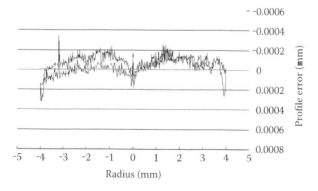

FIGURE 9.61   Profile by SD#1200 after compensation.

TABLE 9.10
Roughness

Wheel	PV (nm)	Ra (nm)
#4000	151.5	9.6
#1200	848.7	17.7
#325	1894.4	109.3

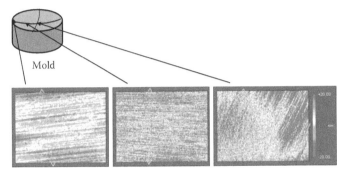

Mold

FIGURE 9.62 Difference in surface quality due to peripheral speed.

TABLE 9.11
Roughness

Wheel	PV (nm)	Ra (nm)
#8000	17.1	1.4
#4000	27.7	3.9
#1200	848.7	17.7
#325	1894.4	109.3

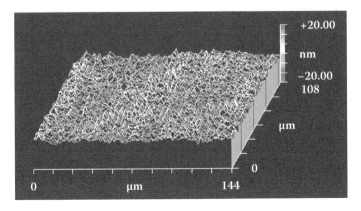

FIGURE 9.63 Roughness patterns by SD#8000.

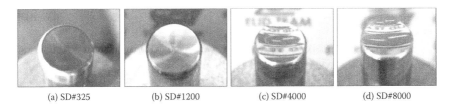

(a) SD#325   (b) SD#1200   (c) SD#4000   (d) SD#8000

FIGURE 9.64 View of ground mold sample on WC. (a) SD#325. (b) SD#1200. (c) SD#4000. (d) SD#8000.

(a) Blank

(b) SD#325

(c) SD#1200

FIGURE 9.65 View of ground mold sample on SiC. (a) Blank. (b) SD#325. (c) SD#1200.

processes with state-of-the-art machining systems under control of ultraprecision parameters.[17–21]

When tools are miniaturized, their surface microstructure and material texture begin to affect their mechanical properties and performances,[17–22] thus undesirably increasing structural sensitivity. In particular, microtools used for micropunching and microcutting require sufficient mechanical strength to withstand the load during machining. If the microtool is poor in surface integrity, a rough surface may act as a fracture origin, degrading the strength of the tool.

With growing demands for new microfabrication technologies capable of ultraprecise and highly accurate machining of hard materials to produce parts, the principal techniques of microgrinding such as ELID are expected to become very important. The construction of desktop fabrication systems and tools based on the concept of applying small processing machines to the fabrication of small parts is therefore considered essential. In our research we have developed several desktop microgrinding machines that provide several dedicated processing functions. This study evaluates the performance of these machines, focusing primarily on a new desktop microtool processing unit for development of fine tools.

9.4.2 FABRICATION OF MICROTOOLS BY **ELID** GRINDING

We performed microtool processing experiments using the machine. One of the developed machines is a three-axis grinding unit. Figure 9.66 shows the external view of the machine developed specifically for microtool fabrication.[23] It holds the material to be machined on the X–Y table that can revolve around the microgrinding spindle. The conditions for the experiment are shown in Table 9.12. Using ELID, small and long shafts as well as microtools could be produced. Figure 9.67 shows a result of fabricating a microtool with a sharp configuration with a column of square cross-section. The tip is extremely narrow, like an order of 2 μm, thus confirming the ability of the machine to produce microtools with a high aspect ratio. It is thought that the sharp corner is required on microtools to perform high-quality cutting or milling. Figure 9.68

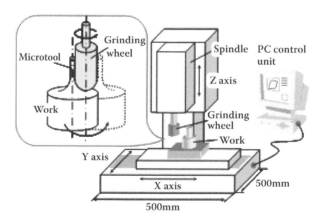

FIGURE 9.66 External view of grinding machine employed for microtool machining.

TABLE 9.12
Experimental Conditions

Workpiece	Cemented carbide alloy
Grinding wheel	Cast-iron bonded diamond wheel (mesh size: #1200, #4000)
Grinding conditions	Wheel rotation: 20,000 min^{-1}
	Depth of cut: 0.5 µm
ELID conditions	Open voltage: 30 V, Peak current: 1 A
	Pulse timing (on/off): 2/2 µs
	Pulse wave: Square

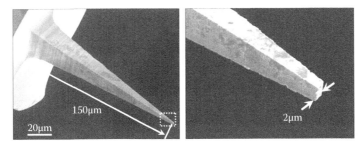

FIGURE 9.67 Microtool grinding result.

shows SEM surface observations of produced microtool edges when two grinding wheel mesh sizes were used of #1200 and #4000. As can be seen in the figure, the finer mesh-sized wheel produced better surface quality and edge sharpness.

A desktop cylindrical grinding machine was also developed to improve processing efficiency. The new unit includes an ELID grinding system and can be used for long cylindrical processing, cutting, boring, polygonal bar processing, and tapering. The unit is not large, as with previous units, but is very compact with exterior dimensions of only 500 mm wide by 500 mm deep by 560 mm high. The most distinguishing feature of this machine is the two opposed grinding wheel heads arranged so as

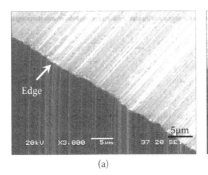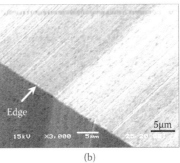

(a)   (b)

FIGURE 9.68 Microphotographs of ground surfaces and edges. (a) #1200. (b) #4000.

FIGURE 9.69 Overview of the desktop cylindrical grinding machine.

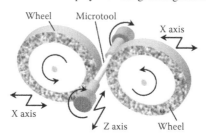

FIGURE 9.70 Schematic diagram of the desktop cylindrical grinding machine.

to pinch the workpiece. The two wheels can grind the workpiece from both directions and can hold the workpiece on two sides to eliminate any deformation of the piece during grinding. This enables highly precise, fine micropins at the end. Each wheel head has its own ELID grinding system. Figure 9.69 shows external views of the unit, and Figure 9.70 shows a schematic diagram. Table 9.13 lists the specifications. The unit has two linear axes, X and Z. A slide mechanism on a cross-roller guide is driven by a step motor to move the mechanism along the linear axes. An indexing unit can be mounted on the main axis section of the workpiece for fabricating a polygon (cross-sectional shape).

Using the new desktop cylindrical grinding machine described in the previous section, we experimentally fabricated a staged cylindrical micropin. Table 9.14 gives

TABLE 9.13

Specifications of the Desktop Cylindrical Grinding Machine

X axis	Positioning accuracy: 0.5 μm
	Grinding spindle speed range: 600–6000 min^{-1}
Z axis	Stroke length: 120 mm
	Spindle speed range; 600–6000 min^{-1}
Dimensions	500 × 500 × 560 mm
Machine weight	80 kg
Power supply	AC100 V

TABLE 9.14

Micropin Processing Conditions

Workpiece	Cemented carbide alloy
Grinding wheel	Cast-iron bonded diamond wheel (mesh size: #1200)
Grinding conditions	Wheel rotation: 3000 min^{-1} Depth of cut: 0.5 μm
ELID conditions	Open voltage: 30 V, Peak current: 1 A Pulse timing (on/off): 2/2 μs Pulse wave: Square

the conditions. The material used was cemented carbide. In preparation for processing, the grinding wheel shape was trued and an ELID initial dressing was performed. Figures 9.71 and 9.72 shows the staged micropins after fabrication. As can be seen in Figure 9.71, a high-aspect staged micropin 4.5 mm in full length and 500 μm at the top was produced. The enlarged photo of the top of the pin as in Figure 9.72 shows a very smoothly finished surface. Because this is a quite smooth surface quality, surface irregularities will not cause fractures during processing as fine as even several microns. Regarding processing efficiency, this system using two grinding wheel heads reduces processing time to about 1/30th that of a conventional system. The configuration of the new grinding unit is very different from conventional grinding

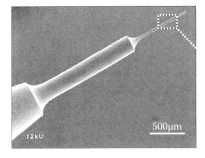

FIGURE 9.71 Staged micropin after processing (#1200).

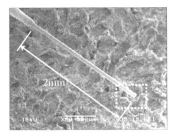

FIGURE 9.72 Staged micropin after processing (#20000).

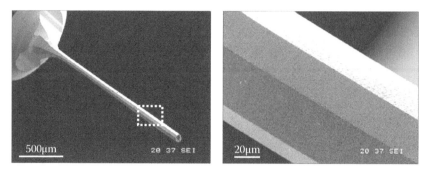

FIGURE 9.73 Polygonal (octagonal) microtools processed with the indexing unit.

machines, making a simple comparison difficult. However, it is obvious that the new machine has excellent processing efficiency. Figures 9.73 and 9.74 shows the polygonal microtools produced using a work indexing unit on the main axis. The results show good corner edges and surface properties.

9.4.3 GENERATION OF MICROTOOLS WITH HIGH-QUALITY SURFACES

Figure 9.75 shows the results of SEM observation of the surface when machining was performed with the grinding wheel mesh size being changed from #325 to #1200, #4000, and then to #20000. The tool was machined as a rectangular shape. As seen in Figure 9.75, increasing the mesh size of the grinding wheel being used from #325 to #20000 produced corresponding improvement in surface quality. Figure 9.76 shows the results of surface roughness measurements by a noncontact-type surface profilometer. The results also indicate that the finer the abrasive size of the grinding wheel, the greater the improvement in surface roughness. In particular, the surface finished with the #20000 grinding wheel, which was attempted for use in this experiment, shows excellent surface characteristics with 1.8 nm in Ra. Figure 9.77 shows the surface characteristics achieved by processing with the #20000 grinding wheel. Grinding marks can be clearly identified, and the figure also indicates that the material was stably removed in ductile mode.

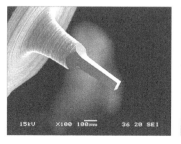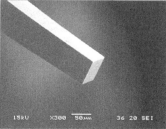

FIGURE 9.74 Polygonal (square) microtools processed with the indexing unit.

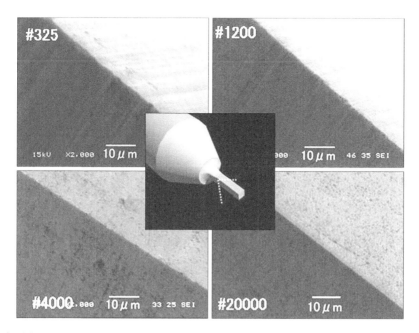

FIGURE 9.75   SEM observation of machined rectangular shape microtools.

When the shape accuracy of the tool is pursued, higher corner-edge accuracy is also very important. In each processed workpiece in Figure 9.75, the shapes of the corner edges were measured with a noncontact-type profilometer. The results are shown in Figure 9.78. From this figure, the corner-edge radius was confirmed to improve with increasing mesh size of the grinding wheel. Thus, the results confirm that microtools could be processed to high-quality ones at nanolevels by the stable performance of both the #20000 grinding wheel having superfine abrasives and ELID.

This section describes the results of testing conducted to investigate rupture strength of the microtool by means of a nanoindentation tester to identify the effects

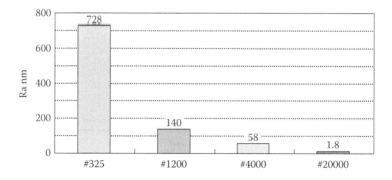

FIGURE 9.76   Surface roughness of finished microtool.

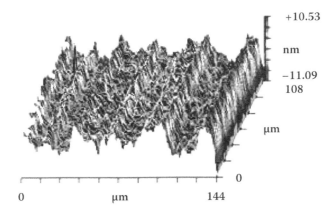

FIGURE 9.77 Surface characteristics achieved by processing with #20000 grinding wheel.

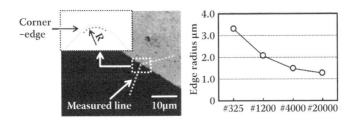

FIGURE 9.78 Measured results of corner-edge radius.

of surface characteristics on mechanical strength. As shown in Figure 9.79, we set a tool in place such that it would produce contact with the indenter at a position 5 μm from the tip of the microtool, and measured indentation load and indentation depth.[22] Figure 9.80 shows the results of nanoindentation testing conducted on microtools finished with grinding wheels of different mesh sizes. As seen in the figure, the smaller the tool diameter, the smaller the rupture strength. Figure 9.81 shows, for the region enclosed by a dotted line (tool diameter: 50 μm) in Figure 9.80, the relationship between surface roughness and load capacity. The tool finished with a grinding

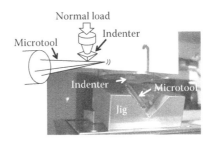

FIGURE 9.79 Overview of microtool rupture strength evaluation system using a nanoindentation tester.

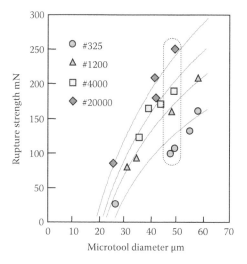

FIGURE 9.80 Results of rupture strength testing.

wheel with finer abrasives clearly has higher durability. These results indicate that differences in surface characteristics exert an extremely significant influence on mechanical strength of the tool.

9.4.4 DEVELOPMENT OF A NEW DESKTOP MICROTOOL MACHINING SYSTEM

The microtool is utilized for various fields in microfabrication. By using microtools as cutting tools, we can fabricate microparts and molds. We have focused on the tool deflection when the spindle is rotating. In this microcutting, deflection of a spindle and detection of a tool tip position have a big influence on processing accuracy or surface quality and abrasion of tool. If chucking of the tool is carried out, deflection certainly arises. Figure 9.82 shows an image during tool rotation (40000 rpm). The diameter of this tool is about 20 μm. The result of having measured deflection in an on-machine measurement device is shown in Figure 9.83. The deflection in the usual chucking is about 6 μm. But it could sometimes be 2 μm and may be 5 μm. It depends on the

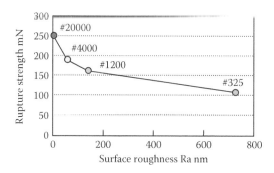

FIGURE 9.81 Relationship between surface roughness and rupture strength.

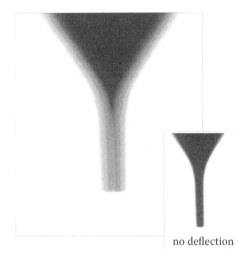

no deflection

FIGURE 9.82   The measurement picture of deflection.

accuracy of the tool or chuck. The tool may be damaged when it is processed in this state using a tool about 10 μm in diameter.

We considered that making tools by an on-machine measurement device should solve these problems. Then, we focused more on measurement devices to combine making and processing tools. A system we have developed that enables one to process with one machine through design to finish will be described next.

Figure 9.84 shows the exterior of the machine developed specifically for microtool machining. The machine is designed to be extremely compact and can be used on a desktop.[24] This machine has three linear axes: X, Y, and Z. Also, an indexing unit

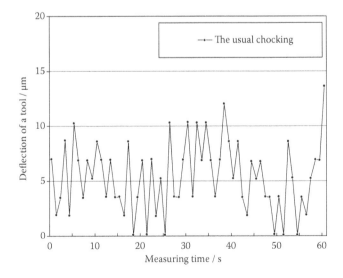

FIGURE 9.83   The deflection measurement result by an on-machine measurement device.

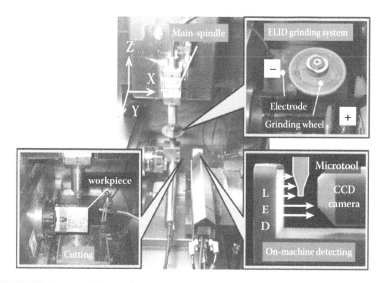

FIGURE 9.84 External view of machine developed for microtool machining.

can be mounted as another rotational axis for workpieces to be fabricated not only into cylindrical shapes but also, if desired, into columnar shapes of square or elliptical cross-sections, or pyramidal shapes having corners. To improve the shape accuracy of desired microtools, an image analysis profilometer (resolution: 0.1 μm) was installed on the machine. Thus, fine-tuning, from an initial diameter of several hundreds of micrometers to a final diameter of several micrometers, has become possible without removal of the microtools from the machine for measurement (Figure 9.85).

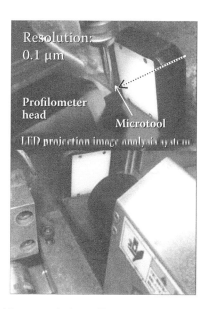

FIGURE 9.85 Installed image analysis profilometer.

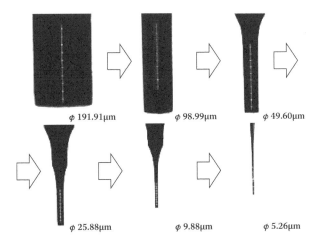

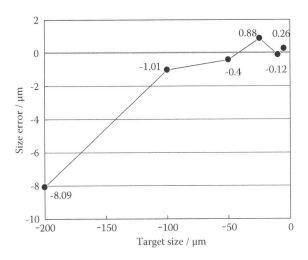

FIGURE 9.86   On-machine measurement picture.

The on-machine measurement device includes a noncontact image pick-up system, a high resolution camera system, and an image processing system. Measurement of a tool diameter and deflection are possible for this device. Moreover, it can be used also for the tool-tip position detection at the time of machining. Feedback processing was performed from the tool of diameter 200 μm to 5 μm. Figure 9.86 shows the pictures of a microtool using an on-machine measurement device. Moreover, the results between target size and size error at micropunching is shown in Figure 9.87. This tells you that it is possible to process with submicrometer precision in comparison with the target size. Moreover, on-machine measurement enables reprocessing

FIGURE 9.87   The target size and size accuracy.

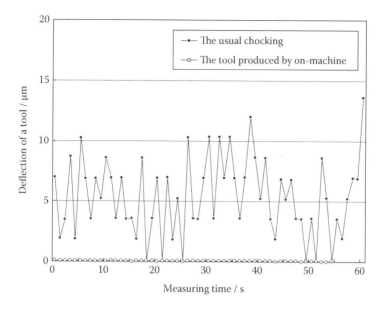

FIGURE 9.88   The deflection measurement result by on-machine measurement device.

from 10 μm to 5 μm. The deflection of the tool in Figure 9.88, in the same figure was about 6 μm. Compared with it, the deflection of an on-machine produced microtool is 0.03 μm (Figure 9.88). Deflection of a microtool can be made very small if a tool is produced in the developed consistent-type system. It is because the portion, which was swaying at the time of chucking, was deleted. Processing conditions could be improved in order not to add excessive power to a tool in processing.

Micro-end-mill was produced to investigate the influence of the deflection exerted on detailed processing. The tool is a one-sheet edge and the diameter is 100 μm. The whetstones used for processing are #325 and #20000. The SEM observation photograph of a micro-end-mill is shown in Figure 9.89. The surface roughness of a tool improves and by using a grinding wheel of #20000 shows that the edge is also sharper. The slot processing experiment was conducted on the tool in the case where tool exchange is performed and after tool production on a plane. SUS316 was used for the work material. Processing conditions are shown in Table 9.15. The SEM observation photograph of a slot processing result is shown in Figure 9.90. The figure shows the microphotograph of a used tool that was produced by the grinding wheel of #325.The processed surface is wavy when there is deflection in a side part. In this case, accuracy and surface roughness worsen. Moreover, this phenomenon may damage a tool. Damage may be given to a tool. On the other hand, a wave is not seen in a case without deflection. That surface roughness is constant and shows good surface quality. The milling result by the tool produced by a grinding wheel of #20000 is shown in Figure 9.91. The edge of tool produced by the grinding wheel of #20000 is sharper than the grinding wheel of #325. Therefore, the surface roughness is better. This factor is so important that a tool becomes small. From these results,

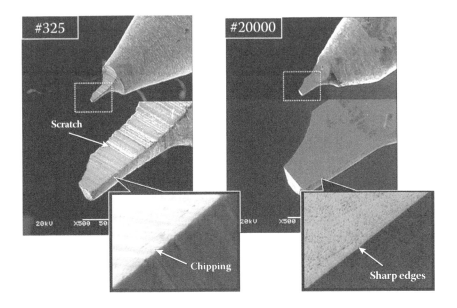

FIGURE 9.89 SEM observation of micro-end-mill.

TABLE 9.15
Cutting Condition

Tool rotational speed	40000 (min^{-1})
Total depth of cut	100 μm
Depth of cut	2 μm/pass
Feed speed	10 mm/min
Processing length (mm)	5 mm/line
Work material	SUS316
Coolant	Kerosene

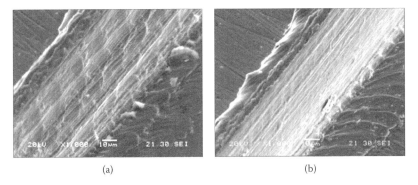

(a)   (b)

FIGURE 9.90 Microphotograph at the side part of machined workpieces. (The used tool was produced by the grinding wheel of #325.) (a) When there is any tool deflection. (b) When there is no tool deflection.

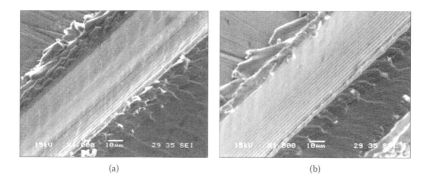

(a)   (b)

FIGURE 9.91 Microphotograph at the side part of machined workpieces. (The used tool was produced by the grinding wheel of #20000.) (a) When there is any tool deflection. (b) When there is no tool deflection.

by reducing deflection of the tool, surface quality of the machined field was able to improve and wear of tool was able to be controlled.

9.4.5 EXAMPLES OF MICROTOOLS FABRICATED BY ELID GRINDING

Controlling the machining conditions enabled a considerable reduction in machining load and enabled machining of distinctive tools like those shown in Figure 9.92, which features smooth, uniform surface characteristics over the entire tool. Figure 9.92a shows an ultrafine tool having a tip diameter of approximately less than 1 μm, while Figure 9.92b shows a tool with extremely sharp corner edges and a large aspect ratio. As described in the previous sections, these extremely fine tools are considered to be made possible by the synergetic effects of the improvement in surface roughness and the diffusion of oxygen atoms.

To determine the effects of the microtool surface characteristics on the practical performance of the tool, we conducted milling and punching tests on stainless steel using microtools manufactured in this study. The results of milling tests are shown in Figure 9.93a. The tests were carried out by punching at a speed of 300 mm/min. For the test, we used a hexagonal shape microtool whose surfaces were finished by a #20000 mesh grinding wheel. Figure 9.93b shows a result of the punching test. The hole has an extremely sharp corner edge and no scratches. In addition, a honeycomb surface structure, shown in Figure 9.93c, was successfully produced by successively making multiple holes with this tool.

9.5 OPTICS FINISHING INTEGRATING ELID GRINDING AND MRF

ELID grinding was proposed by one of the authors for automatic dressing of the grinding wheel while performing grinding for a long time. It is effective and has been widely used for grinding hard and brittle optical materials. However, those surfaces produced by fixed abrasive grinding are characterized by considerable subsurface damage and microcracks. Those damages should be typically removed

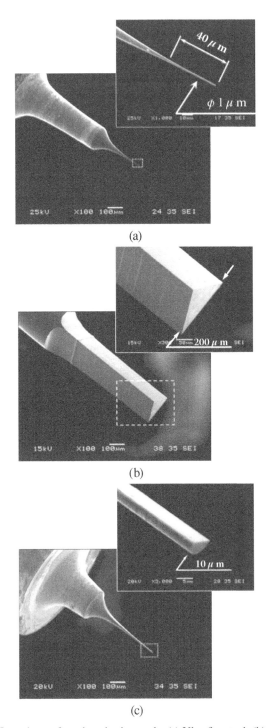

FIGURE 9.92 Overviews of produced microtools. (a) Ultrafine tool. (b) Hexagonal micro-tool (c) Half-moon shape. (d) Triangular shape. (e) Milling tool shape.

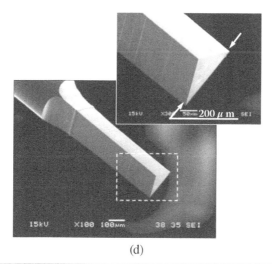

(d)

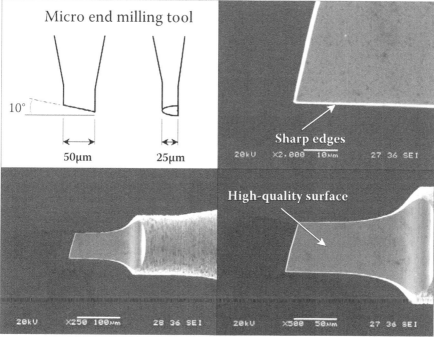

(e)

FIGURE 9.92 (Continued).

by loose abrasive grinding followed by polishing with a friable abrasive. Magneto-rheological finishing (MRF) is a novel precision finishing process for deterministic form correction and polishing of optical materials by utilizing magneto-rheological fluid. In this chapter, a nanoprecision synergistic finishing process integrating MRF and ELID grinding was proposed to shorten the total finishing time and improve

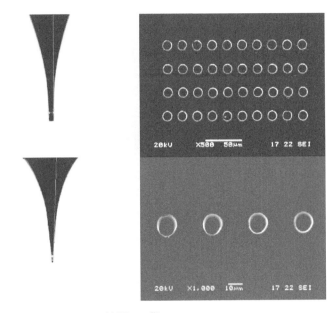

(a) Micromilling tests

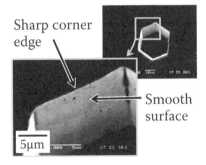

Sharp corner edge

Smooth surface

5μm

(b) Punched hole

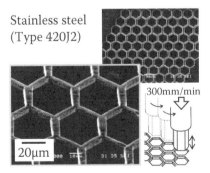

Stainless steel (Type 420J2)

300mm/min

20μm

(c) Punched honeycomb structure

FIGURE 9.93 Results of milling and punching test. (a) Micromilling tests. (b) Punched hole. (c) Punched honeycomb structure.

finishing quality. Many nanoprecision experiments have been carried out to grind and finish some optical materials such as glass, silicon, and silicon carbide. ELID grinding, as prefinishing, is employed to obtain high efficiency and high surface quality, and then, MRF, as the final finishing, is employed for further improved surface roughness and form accuracy. In general, form accuracy of $\lambda/10$ to $\lambda/20$ nm peak-to-valley (PV) and surface microroughness less than 10 angstrom were produced in high efficiency.

9.5.1 INTRODUCTION

An automatic and high-efficiency nanoprecision manufacturing technique for an optical lens surface with subnanometer root mean square (RMS) surface microroughness, ~20 nm figure accuracy, and damage free is greatly demanded by advanced optical fields. A main conventional manufacturing process of an optical lens consists of CNC grinding and polishing.

Conventionally, a typical polishing tool was approximately the same size as the lens surface and was the perfect mate for the lens, cushioned by a slurry of water and soft polishing abrasives. The two surfaces were rotated and oscillated against each other to yield a very smooth optical surface. It was easy to achieve a $\lambda/4$ PV figure with a 1 nm RMS finished surface. So far, a computer-controlled subaperture polishing processing technology had been developed. A mechanically flexible used polishing tool was smaller than the workpiece, and even actively controlled reduces misfit.[25,26] However, those approaches suffered from the same problem: they were not deterministic because the polishing tools change uncontrollably with time, owing to nonuniformity of loose abrasive distribution and polishing pressure distribution, and wear of the polishing pad. Those polishing methods have great difficulty in satisfying these demands.

In conventional grinding processes, problems such as wheel loading and glazing can be encountered while grinding a glass lens with fine abrasive wheels.[27–30] Therefore, we proposed ELID grinding. ELID is an efficient method to dress the grinding wheel while performing grinding for a long time. Effective dressing is possible even if the diamond grains on the wheel surface are very small in size. Grinding optical elements using the ELID technique is a viable alternative and effective.[31–36] However, in grinding, the surface produced by fixed abrasive grinding of these optical materials was characterized by considerable subsurface damage and microcracks on the order of the abrasive grain size. This damage should be typically removed by polishing with a friable abrasive such as cerium or aluminum oxide.[37,38] However, the figure accuracy built into the original ground surface was often compromised as a result of this conventional polishing operation. Therefore, for realizing the extreme precision and surface microfinish, we sought the best polishing method.

A revolutionary finishing method, MRF, was invented and developed by an international group of collaborators and commercialized by QED Technologies Inc. in 1997.[39–43] MRF is based on a magnetorheological (MR) fluid. In MRF, a normal force of the order of 0.01 N between the abrasive particle and the parts is the key to removal in most classical polishing processes. Shear stress in the converging gap and the lateral motion of polishing abrasives across the part surface raises material removal without surface damage, leaving the part extremely clean, pit, and the surface scratch-free. This has been shown to minimize the embedding of polishing powders. It is suggested that MRF is an excellent candidate for polishing optics elements. MRF is deterministic because the polishing tool removal rate changes little, and the removal function is interferometrically characterized. If the removal is accurately known, the computer algorithms function to high precision. Other advantages are that the polishing tool is easily adjusted and conforms perfectly to the workpiece surface, enabling aspheric polishing.[44–50]

So, ELID grinding was used as a high-precision and high-efficiency grinding method for hard and brittle materials by using in-process electrolytic dressing. MRF was a novel precision finishing process for deterministic figure correction and polishing of optical material. In this section, an ultraprecision synergistic finishing process of ELID grinding and MRF is developed and applied to the fabrication of a glass lens, SiC, silicon mirror, and so forth. As prepolishing, in ELID grinding, good surface quality may be obtained in high efficiency, and then, the last high form accuracy and surface roughness may be obtained by MRF. Surface accuracy of ~20 nm peak-to-valley and surface microroughness less than 10 angstrom are produced in high efficiency. Research results have shown that this new synergistic finishing process may obtain both high efficiency and high quality, and may be used to replace conventional optics manufacturing processes.

9.5.2 REMOVAL RATES OF DIFFERENT MATERIALS BY MRF

9.5.2.1 Removal Rate of BK7 Glass

Figure 9.94 shows the removal "spot" for a 25 mm diameter, 60 mm radius of a curvature BK7 glass lens, immersed in the MR suspension for 2 seconds (conditions described in Table 9.16). The spot area size is approximately 3.3 × 2.3 mm. The maximum depth of the spot is 181 nm, the peak removal rate may be calculated to be about 5.43 μm/min, and the volume removal rate is 0.0144 mm³/min.

9.5.2.2 Removal Rate of SiC

Figure 9.95 shows the removal spot for a flat surface of 30 mm diameter when the SiC parts are immersed in the MR suspension for 20 seconds. The MRF conditions are described in Table 9.17. The mean depth of the removal spot was 303 nm, and

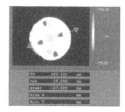 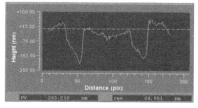

FIGURE 9.94 MRF removal function on BK7 after 2 seconds.

TABLE 9.16
MRF Conditions of BK Glass

MRF machine	Q22Y MRF system (QED)
Finishing mode	Uniform removal and figure correction
Workpiece	No. 1—Materials: Bk7 glass, F.A. 25 mm, concave surface, curvature radius R60 mm
	No.2—Materials: Bk13 glass, F.A. 70 mm, plane
Delivery system setting	Magnet: 12 A; Wheel revolution: 350 rpm; Suction pump revolution: 110 rpm
	Centrifugal pump revolution: 2800 rpm
Delivery system reading	Pressure: 7.08 PSI; Flow rate: 0.2 l/m; Viscosity: 38.5 cP
Ribbon parameters	Wheel: 50 mm; Ribbon height: 1.31 mm; Gap: 1.07 mm; Depth: 0.24 mm
MR fluid	Cerium oxide abrasive

the peak removal rate and the volume removal rate may be calculated to be about 0.909 μm/min and 0.005 mm³/min, respectively.

9.5.2.3　Removal Rate of Silicon Mirror

Figure 9.96 shows the removal spot for silicon mirror, immersed in the MR suspension for 3 seconds. Table 9.18 shows the MRF conditions. The peak removal rate may be calculated to be about 5.04 μm/min and the volume removal rate is 0.036 mm³/min.

9.5.3　Finishing Characteristics of ELID and MRF

In general, a finishing process may be classifed as rough finishing and mirror finishing. MRF should be positioned in the final finishing. It can obtain very good finishing quality. However, it uses MR fluid as tools, and its removal rate is lower than grinding, So, a prefinished surface with a better surface from accuracy is demanded, for example, 4λ-λ PV. Although it can also remove the larger PV error, it is time consuming. To obtain the best polishing quality at the lowest cost and much less time, we think that using ELID grinding, obtaining λ/2~λ/5 PV, and then using MRF, improves it to λ/10~λ/20 PV; this is the best alternative.

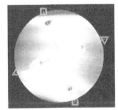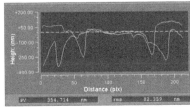

FIGURE 9.95　MRF removal function on SiC after 20 seconds.

TABLE 9.17
MRF Conditions of SiC Plane

MRF machine	Q22Y MRF system (QED)
Finishing mode	Uniform removal
Workpiece	Materials: SiC
	Size: Diameter Φ30 mm, thickness: 10 mm
	Plano surface, before finishing: 0.95 μm RMS
Delivery system setting	Magnet: 12 A; Wheel rotation: 450 rpm; Suction pump rotation:
	120 rpm; Centrifugal pump rotation: 3000 rpm
Delivery system reading	Pressure: 7.8 PSI; Flow rate: 0.2 l/m; Viscosity: 43.2 cP
Ribbon parameters	Ribbon height: 0.9 mm; Gap: 0.55 mm; Depth: 0.35 mm
MR fluid	D10(QED), diamond abrasives

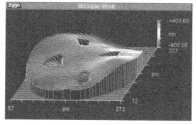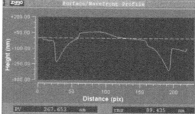

FIGURE 9.96 Optical roughness map of surface after taking spots.

TABLE 9.18
MRF Conditions

MRF machine	Q22Y MRF system (QED)
Finishing mode	Uniform removal and figure correction
Workpiece	Material: Silicon; Diameter: Φ20 mm
Delivery system reading	Pressure 6.6 PSI; Flow rate: 0.2 l/m; Viscosity: 35.5 cP
Ribbon parameters	Wheel: 50 mm; Ribbon height:0.97 mm; Gap: 0.66 mm; Depth 0.33 mm
MR fluid	Diamond abrasive
Finishing time	1.19 min + 1.13 min (2 cycles)

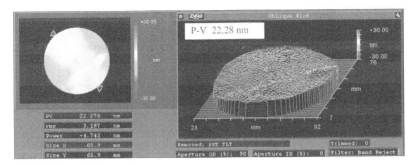

FIGURE 9.97   Surface map after MRF for BK13.

Using the nanoprecision synergistic finishing process integrating MRF and ELID grinding, high efficiency and higher surface quality of optical materials such as glass, silicon, and silicon carbide can be obtained. In general, form accuracy of ~20 nm PV and surface microroughness less than 10 angstrom are produced in high efficiency.

9.5.3.1   Form Accuracy of ELID and MRF

9.5.3.1.1   Form Accuracy of BK13 Plane Surface

For clarifying the form corrective effective of MRF, a BK13 plane blank glass is used to test the integrated ELID grinding using #4000 wheel and MRF. Table 9.19 shows the ELID grinding conditions. The MRF conditions are shown in Table 9.16, and the lens is No. 2. The finished surface form is appreciated by a Zygo interferometer. After ELID grinding, $\lambda/5$ PV ($\lambda = 632.8$ nm) is obtained. And then, after MRF for 30 minutes, the form accuracy is improved to $\lambda/28$ (22.28 nm) PV (Figure 9.97). From this result we know that the figure corrective ability is extremely high.

TABLE 9.19
ELID Grinding Conditions

Workpiece	SiC blank, Size: $\varphi30$ mm $\times$ 10 mm
Grinding machine	Horizontal rotary grinding machine HSG-10A2
Grinding wheel	CIB diamond wheel, $\varphi60 \times W5$ mm, #4000 M100
	Metal-resin bonded wheel #12000 M75
Grinding conditions	Wheel rotational speed: 2000 min^{-1}; Workpiece
	rotational speed: 300 min^{-1}; Feed rate: 1 μm/min
	Total depth of cut: 20 μm for #4000, 0.5 μm for #12000.
ELID conditions	Open circuit voltage (Eo): 90 V; Peak current (Ip): 10 A;
	On/off time (τ_{on}/τ_{off}): 2/2 μs; Pulse wave: Square
Grinding fluid	NX-CL-CM2 (2% dilution of water)

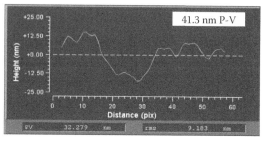

(a) Before MRF (After ELID grinding)

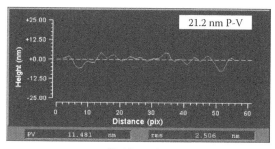

(b) After MRF

FIGURE 9.98   Figure accuracy (a) before MRF (after ELID grinding) and (b) after MRF.

9.5.3.1.2   Form Accuracy of SiC Plane Surface

Figure 9.98 shows a finishing example of an SiC plane surface by integrating ELID grinding and MRF. After ELID grinding using a #4000 wheel, a surface of 43 nm PV is obtained. After MRF for 3 minutes, the surface can be improved to 21.2 nm PV. As shown in Figure 9.99, the peak-to-valley of surface produced by #4000 and #12000 wheels and MRF (diamond grit size 1.4 μm) are 12.7 nm, 2.7 nm, and 1.3 nm in order. Although the diamond abrasive grit size in the MRF tool is near to that of the #12000 grinding wheel, MRF demonstrates a smaller cutting depth. This is because the abrasive removes materials by shear stress in the converging gap and the lateral motion of polishing abrasives across the part surface; it does not force abrasives to cut into the materials as in a fixed abrasive grinding mode. Therefore, MRF may lead a cleaner pit and scratch-free surfaces without surface damage.

9.5.3.1.3   Form Accuracy of Curvature BK7 Lens

To verify obtainable form precision and surface microroughness by using the synergistic finishing process, a 25 mm diameter and 60 mm radius of curvature BK7 glass lens is tested. At first, ELID grinding is conducted by using a #1200 wheel, and glass material is removed mainly in brittle fracture mode. Surface form accuracy λ/5 PV is obtained. Then, MRF is conducted for 17 minutes (two cycles), using removal spot as shown in Figure 9.94. MRF conditions are shown in Table 9.16. The finished surface form is measured by a Zygo interferometer, and the form accuracy is improved to λ/18 PV (Figure 9.100) in a short time.

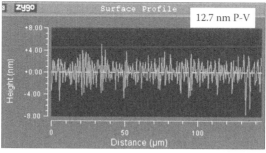

(a) After grinding by #4000

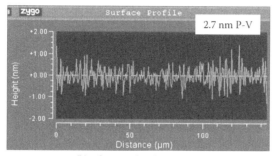

(b) After grinding by #12000

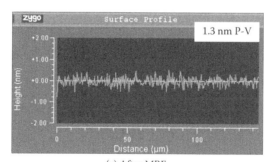

(c) After MRF

FIGURE 9.99 A comparison of microcutting mark of generated surface. (a) After grinding by #4000 wheel. (b) After grinding by #12000 wheel. (c) After MRF.

9.5.3.1.4   Form Accuracy of Silicon Wafer

Figure 9.101 shows the form accuracy of silicon wafer materials by ELID grinding and MRF. The initial form accuracy is 208.5 nm PV after ELID grinding in Figure 9.101a. After the first cycle of MRF, the form accuracy is improved to 34.8 nm PV in Figure 9.101b; after the second cycle of MRF, the form accuracy is further improved to 24.5 nm PV in Figure 9.101c. From this result we know that the figure corrective ability is high. It is seen that extremely high form accuracy could be realized simultaneously in a high efficiency.

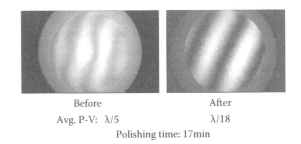

Before   After
Avg. P-V: λ/5   λ/18
Polishing time: 17min

FIGURE 9.100   Figure accuracies before and after MRF for glass lens.

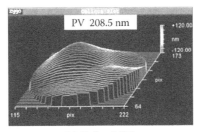

(a) Before MRF

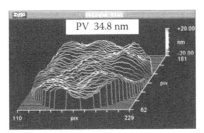

(b) After 1st cycle of MRF

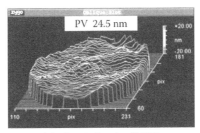

(c) After 2nd cycle of MRF

FIGURE 9.101   Figure map of surface. (a) Before MRF. (b) After first cycle of MRF. (c) After second cycle of MRF.

9.5.3.1.5   Form Accuracy of Silicon Mirror

A concave silicon part (full aperture 40 mm, clear aperture 35 mm in diameter, and 1800 mm radius of curvature) is generated by ELID grinding using a #4000 wheel in an ultraprecision machining system. The ground surface by a #4000 wheel is measured with an on-machine measure system (the Nexsys Corp.). Before grinding, the initial ground surface is measured for calculating the distribution of the form deviation profiles. With these deviation profiles, the NC data is compensated. Form accuracy is 0.3114 μm PV after ELID grinding by a #4000 wheel (Figure 9.102a), and then it is improved remarkably to 70 nm PV after MRF (Figure 9.102b).

9.5.3.2   Microsurface Roughness of ELID and MRF

9.5.3.2.1   Surface Roughness of BK7 Glass Sphere

To verify obtainable surface microroughness using the synergistic finishing process, the same BK7 glass lens and same conditions are used in a test. At first, ELID grinding is conducted using a #1200 wheel, and a surface microroughness of 261 nm RMS is obtained (Figure 9.103a). Then, MRF is conducted for 17 minutes (two cycles). As a result, surface microroughness of 0.56 nm RMS is improved considerably in Figure 9.103b. In a short time, the glass lens surface could be finished to subnanometer surface microroughness.

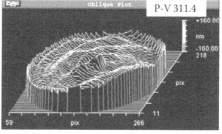

(a) after ELID grinding by #4000 wheel

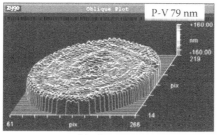

(b) after MRF

FIGURE 9.102   Form map of surface before and after MRF. (a) After ELID grinding by #4000 wheel. (b) After MRF.

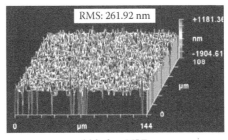

(a) before MRF

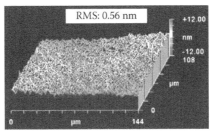

(b) after MRF

FIGURE 9.103 Optical roughness map of surface before and after MRF. (a) Before MRF. (b) After MRF.

9.5.3.2.2 Surface Roughness of CVD-SiC

In this experiment, the conditions of ELID grinding are shown in Table 9.19 and MRF conditions are shown in Table 9.17. A same rough surface in Figure 9.104a and Figure 9.104b was ground by #4000 and #12000 wheels, and then is finished by MRF as shown in Figure 9.104c. Those surfaces generated by #4000 and #12000 wheels and MR fluid are evaluated comparatively. The obtainable microroughness by #4000 and #12000 wheels and MR fluid are 5.8 nm, 3.0 nm, 2.5 nm, respectively. Compared to the #12000 ground surface, the microroughness of the finished surface by MR fluid is not considerably improved. The 2–3 nm RMS value may be considered to be the limited surface-roughness in these finishing conditions.

The finishing operation starts from a very coarse surface with 939 nm RMS (Figure 9.105a). After MRF for 18.5 hours, the surface microroughness 2.8 nm RMS is obtained (Figure 9.105b). The SiC surface could be finished to nanometer surface microroughness. If using ELID grinding for 30 minutes, the roughness may be decreased to 5.7 nm RMS, and then using MRF for 3 minutes, down to 2.5 nm. It is seen that the integrated process shortens finishing time.

9.5.3.2.3 Surface Roughness of Silicon Wafer

Plano silicon parts were generated by ELID grinding in an ultraprecision machining system. As shown in Figure 9.106a, ground surface by a #4000 wheel, generated in a ductile mode, a surface microroughness less than 4.9 nm RMS was

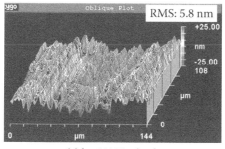

(a) by #4000 wheel

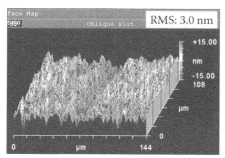

(b) by #12000 wheel

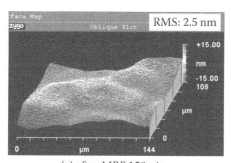

(c) after MRF 150min

FIGURE 9.104 Comparison of optical roughness map of generated suface. (a) By #4000 wheel. (b) By #12000 wheel. (c) After MRF for 150 minutes.

obtained. After MRF, the surface roughness was improved to 0.6 nm RMS (Figure 9.106b).

9.5.3.2.4   Surface Roughness of Fused Silica Glass

A concave silicon part (full aperture 40 mm, clear aperture 35 mm in diameter, and 1800 mm radius of curvature) was generated by ELID grinding using a #4000 wheel in an ultraprecision machining system. Grinding and ELID conditions are

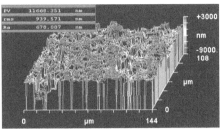

(a) Before MRF RMS: 939.6 nm

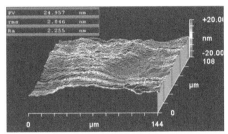

(b) After MRF   RMS: 2.8 nm

FIGURE 9.105 Optical roughness map of surface before and after MRF. (a) Before MRF RMS: 939.6 nm. (b) After MRF RMS: 2.8 nm.

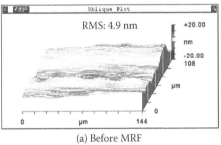

(a) Before MRF

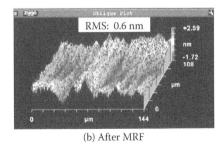

(b) After MRF

FIGURE 9.106 Optical roughness map of surface. (a) Before MRF. (b) After MRF.

TABLE 9.20
ELID Grinding Conditions of Fused Silica Glass

Grinding machine	Ultraprecision mirror surface 4-axis grinding system with 1 nm resolution
Grinding wheel	CIB diamond wheel, φ60 × W5 mm, #4000 straight type, Concentration: 75
Grinding conditions	Wheel rotational speed: 3000 min^{-1}; Workpiece rotational speed: 500 min^{-1}; Feed rate: 5 mm/min; Depth of cut: 1–5 μm/pass
ELID conditions	Open circuit voltage (Eo): 80 V; Peak current (Ip): 10A; On/off time (τ_{on}/τ_{off}): 2/2 μs; Pulse wave: Square
Workpiece	Material: Fused silica glass; Size: 315 mm × 251 mm
Grinding fluid	NX-CL-CM2 (2% dilution of water)

shown in Table 9.20. The grinding wheel was ELID trued by #140 truing wheel. In Figure 9.107, ground surface by #4000, surface microroughness less than 5.16 nm RMS was obtained, and was improved to 0.75 nm RMS.

To verify the relations between a prefinished surface and MRF, the following experiments were conducted. A concave fused silica part (50 mm in diameter, 500 mm radius of curvature) was generated by ELID-grinding in an arc-enveloped grinding system. Castiron bonded diamond wheels (#1200, #4000) were used. In Figure 9.108a,

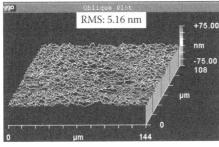

(a) Before MRF

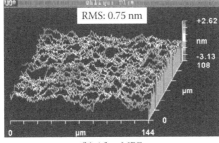

(b) After MRF

FIGURE 9.107   Optical roughness map of surface. (a) Before MRF. (b) After MRF.

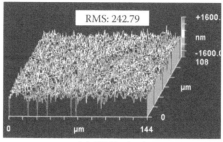

(a) #1200 wheel

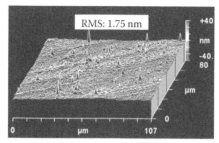

(b) #1200 wheel + MRF 50 min

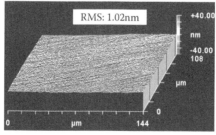

(c) #1200 wheel + MRF 80 min

FIGURE 9.108   After ELID grinding by #1200 wheel and after MRF. (a) #1200 wheel. (b) #1200 wheel + MRF 50 min. (c) #1200 wheel + MRF 80 min.

ground surface by a #1200 wheel is generated mainly in brittle fracture mode, and microroughness is obtained to 242.79 nm RMS. On the other hand, ground surface by a #4000 wheel, generated in ductile mode, surface microroughness less than 69.29 nm RMS is obtained (Figure 9.109a). The surface roughness strongly depends upon the grinding mode. Thereafter, two ground surfaces by #1200 and #4000 were polished by MRF separately; MRF conditions are shown in Table 9.21. For the #1200 surface, after MRF for 50 minutes, surface roughness was improved to 1.75 nm RMS (Figure 9.108b). However, after MRF for 80 minutes, surface roughness was improved considerably to 1.02 nm RMS (Figure 9.108c). In Figure 9.109b, for #4000 surface after MRF 50 minutes, the surface roughness was improved to 0.98 nm RMS from before polishing 69.29 nm RMS. It is seen that the MRF time was shorter in polishing #4000 than #1200.

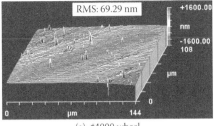

(a) #4000 wheel

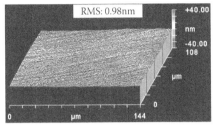

(b) #4000 wheel + MRF 50min

FIGURE 9.109 After ELID grinding by #1200 wheel and after MRF. (a) #4000 wheel. (b) #4000 wheel + MRF 50 min.

9.5.4 CONCLUSIONS

Owing to a lower removal rate of MRF for ultrahard material, MRF is time consuming and has low efficiency for polishing a rough optical surface. Its prefinished surfaces should have better figure accuracy and better surface roughness; ELID grinding may meet those needs, so ELID is the best alternative and can decrease finishing time. ELID grinding was conducted before MRF. By applying the ultraprecision synergistic finishing process of ELID grinding and MRF, surface accuracy of the $\lambda/10 \sim \lambda/20$ nm peak-to-valley and surface microroughness less than 10 angstrom for glass lens were successfully produced in high efficiency. Enabling synergistic technologies may shorten finishing time and eliminate the industry's reliance on the

TABLE 9.21
MRF Conditions

Workpiece	Materials: Fused silica; Size: Full aperture: Φ50 mm; Curvature radius: R500 mm, concave surface
Delivery system setting	Magnet: 15 A; Wheel revolution: 380 rpm
	Suction pump revolution: 120 rpm;
	Centrifugal pump revolution: 2800 rpm
Delivery system reading	Pressure: 7.08 PSI; Flow rate: 0.2 l/m; Viscosity: 32.5 cP
Ribbon parameters	Ribbon height: 1.21 mm; Gap: 1.01 mm; Depth: 0.20 mm
MR fluid	Cerium oxide abrasive

specialized skills required to operate today's costly and labor-intensive conventional manufacturing processes. Furthermore, these technologies extend the manufacturing state-of-the-art and provide optical manufacturers with the cost-effective capability to produce optics glass and lens.

9.6   SURFACE MODIFICATION

As a potential future application of ELID, we have developed a new surface-modifying fabrication process that generates a thin oxide layer on the surface of materials by controlling the fabrication atmosphere during ultraprecision mechanical fabrication processes.

Machining processes used to produce a surface are influenced by mechanical, thermal, and chemical loadings in the contact area. When using a grinding process, the physical and chemical characteristics of the surface layer after processing differ from those of the base material of the substrate because of the effects of elastic–plastic deformation and heat generation resulting in abrasives rubbing the surface of a workpiece. Considerable progress has been made in the study of phenomena that occur in the surface and subsurface after the surfaces have been processed, resulting in typical states such as grinding burns, grinding cracks, work-affected layers, and changes in hardness and residual stress.[51,52] However, little research has been done into surface functionalities after processing. In a past study, for instance, it was found that residual stress in surfaces enhances fatigue strength;[53] however, many points still remain unexplained, and they represent important issues for future research.

In particular, there have been recent rapid advancements in the development of high-precision grinding using superabrasive wheels.[54] It has been reported that a mirror-finished surface produced by high-precision grinding has exceptional functionalities since it has a surface layer that has been modified rather than damaged by processing.[65]

This method is expected to produce various effects such as inducing an oxidization reaction on an active new surface immediately after processing, to thereby coat the surface with an anticorrosion film (I in Figure 9.110); advantageously transforming the components of the tool or grinding fluid to the surface (II in Figure 9.110); and intentionally creating an interface to improve adhesive strength with the coating

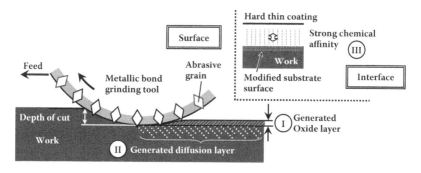

FIGURE 9.110   Concept of surface and interface functional modification.

layer (III in Figure 9.110). Reactions I and II control the surface layer of the workpiece, and reaction III controls the interface between the workpiece and the coating layer. For all of these reactions, nanometric tribochemical reactions are positively utilized during grinding. This section discusses applications of these latest techniques in detail by introducing two cases.

9.6.1 SURFACE MODIFICATION FOR STAINLESS STEELS

We selected a type of stainless steel that is anticipated to have a wide variety of industrial applications as workpieces and conducted a detailed analysis of the surfaces after they had been mirror-finished by high-precision grinding. Furthermore, on the assumption that stainless steel will be adopted for precision molds, we evaluated its properties, including its mechanical properties, corrosion resistance, and high-temperature oxidation resistance, which are necessary surface functionalities for molds.

9.6.1.1 Proposed Modification Technique and Experimental Procedures

Figure 9.111 shows the proposed grinding system and a schematic illustration of surface modification reactions. In the region labeled A, an electrode with a gap of about 0.1 mm was placed directly opposite to a conducting metal-bonded grinding wheel, and positive (+) and negative (−) potentials were respectively applied. Applying a current during grinding via an alkaline conductive grinding fluid electrolyzes the bonding material of the wheel, so that the continuous protrudent abrasive of the grinding wheel can be reasonably maintained, enabling a stable mirror-finish grinding.[54]

Furthermore, in the region labeled A, there is a mechanism that induces water electrolysis and functions to dramatically increase the concentrations of hydroxide ions and of dissolved oxygen in the grinding fluid. By applying a weak positive potential to the workpiece, large quantities of the hydroxide ions in the grinding fluid

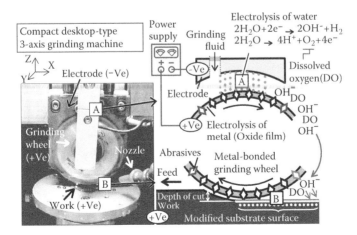

FIGURE 9.111 Overview of the proposed grinding system and a schematic illustration of surface modification reactions.

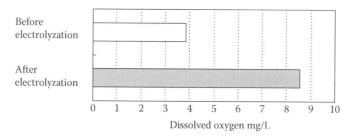

FIGURE 9.112   Concentrations of dissolved oxygen.

are actively attracted to the newly produced surfaces of the workpiece by electrophoresis, and they form a stable oxide layer by an anodic oxidation reaction. Moreover, the dissolved oxygen in the grinding fluid penetrates and diffuses into the substrate of the workpiece. Figure 9.112 shows the result of measuring the concentrations of dissolved oxygen in the vicinity of region A before and after electrolyzation. The concentration of dissolved oxygen is significantly increased after electrolyzation.

On the other hand, at the contact zone (indicated by B in Figure 9.111) the abrasive elements are expected to penetrate and diffuse into the substrate of the workpiece because of a tribochemical reaction between abrasives and workpieces (a solid-state reaction of abrasive elements).

Stainless steel (type 420J2) was used as the material to which mirror finishing and surface modification were applied by using the aforementioned processing system. Table 9.22 shows the processing conditions. The grinding wheel abrasives were selected so as to control the modifying process. Metal-resinoid hybrid bonded grinding wheels (#8000, having approximately 2 μm diameter abrasives) with three different abrasives (diamond [C], silicon dioxide [SiO_2], and alumina [Al_2O_3]) were used for finishing. These specimens are referred to as the C-series, Si-series, and Al-series, respectively. For comparison, another workpiece was polished using #4000 emery paper and then buffed with alumina abrasives prior to processing with the proposed system. This procedure is hereafter referred to as the P-series (polish). Table 9.23 lists the equipment utilized for surface analysis and functionality evaluation of the processed workpieces.

TABLE 9.22
Experimental Conditions

Workpiece	Type 420J2 stainless steel
Machine	Compact desktop-type three-axis grinding machine
Grinding wheel	#8000 metal-resinoid hybrid bond diamond wheel (C-series), SiO_2 wheel (Si-series), Al_2O_3 wheel (Al-series)
Grinding fluid	Chemical solution type grinding fluid (5% dilution to water)
Grinding conditions	Wheel rotation: 1000 min⁻¹; Feed rate: 200 mm/min; Depth of cut: 0.5 μm
Electrical conditions	Open voltage: 70 V; Peak current: 1A; Pulse timing (on/off): 2/2 μs; Pulse wave: Square

TABLE 9.23

Lists of Analyzers Used in This Study

Surface Analysis

Contact type surface profilometer, glow discharge optical emission spectroscopy (GD-OES), x-ray photoelectron spectroscopy (XPS), Nanoindentation testing, x-ray diffraction stress analyzer, surface potential measurement instrument

Surface Functionality Evaluation

Pin-on-disc tribology system, plane bending fatigue testing system, anodic polarization corrosion test system, ionized deposition coating system, microscratching test system, contact angle measurement system

9.6.1.2   Analysis of Mirror-Finished Surfaces

Surface roughness. The finishing process was performed for the aforementioned three types of #8000 grinding wheels and the surface roughness were measured using a contact-type surface profilometer. The surface roughness Ra values were about the same and ranged from 8 to 10 nm for all abrasives.

Analysis of oxygen element. Figure 9.113 shows the results of measuring the diffusion of elemental oxygen in each series, measured from the surface in the depth direction using glow discharge optical emission spectroscopy (GD-OES). The figure shows that all series have a higher concentration of oxygen in the vicinity of the surface than the P-series, and also a sufficient amount of penetration into the interior of the substrates. We consider that this result can be attributed to the effect described in the previous section in which hydroxide ions and dissolved oxygen penetrate and diffuse into the workpiece surfaces activated by processing.

Analysis of abrasive elements. Focusing on the diffusion of the abrasive elements, an elemental analysis was conducted on the Al-series by means of x-ray photoelectron spectroscopy (XPS). Figure 9.114 shows the elemental analysis profile in the depth direction. This figure clearly confirms the diffusion of elemental Al from the surface

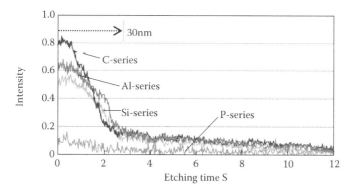

FIGURE 9.113   Oxygen elements intensity profile in the depth direction analyzed by using GD-OES.

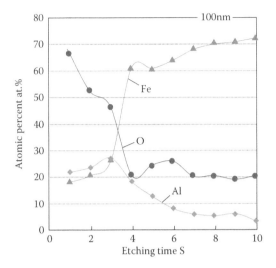

FIGURE 9.114 Elemental analysis profile of the Al-series.

to a depth of about 30 nm. The results suggest that a phenomenon occurred during the grinding process in which the abrasives penetrated and diffused into the workpiece surfaces at the machining contact points. Furthermore, from the elemental spectra obtained, Figure 9.115 shows the results of the spectra of the peak in the vicinity of the aluminum binding energy (72.9 eV). This figure shows that this peak has shifted to a higher energy relative to the aluminum binding energy. This finding indicates that penetration and diffusion by elemental aluminum enhances the oxidative tendency.

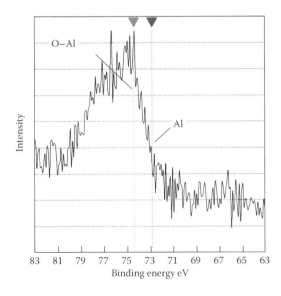

FIGURE 9.115 XPS results for analysis of aluminum.

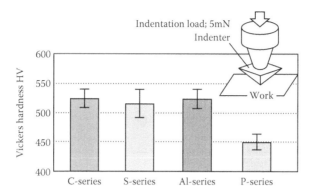

FIGURE 9.116   Results of nanoindentation test.

Although this section describes only the diffusion of elemental aluminum on the Al-series, elemental C and Si diffuse into surfaces processed by the C- and Si-series, as has been previously reported.[55] These results raise the possibility of controlling elements to be diffused into processed surfaces by selecting abrasive elements used for grinding wheels.

9.6.1.3   Surface Functionality Evaluation

Evaluation of mechanical properties: Hardness of top surface layer. For stainless steel used for molds, the mechanical strength properties are a significant surface functionality. In this section, nanoindentation testing was used to evaluate the hardness of the processed and modified top surface layer. The results are shown in Figure 9.116. The hardness of the P-series was 450 HV, whereas the other three workpieces had significantly higher hardness of about 530 HV.

Residual stress measurements. Figure 9.117 shows the results of measuring the residual stress of the surfaces in the C- and P-series. This figure shows that the

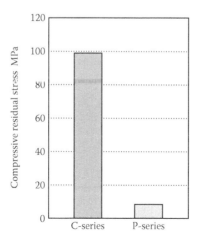

FIGURE 9.117   Comparison of the compressive residual stress.

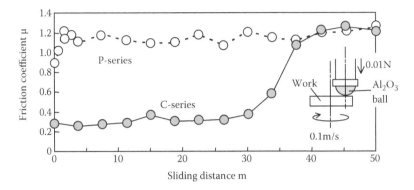

FIGURE 9.118 Results of sliding friction test.

C-series has a compressive residual stress of about 100 MPa, whereas the P-series finished by polishing had very little residual stress.

Tribological properties. Figure 9.118 shows the results of sliding friction testing carried out on the C- and P-series. A pin-on-disc tester was used for these measurements. Although the friction coefficient is dependent on the test conditions, the initial friction coefficient of the C-series is approximately a quarter that of P-series. This enhanced hardness of the modified surface is considered to be one factor that contributes to the improvement in the tribological characteristics.

Fatigue properties. Figure 9.119 shows an S–N diagram in which the stress amplitude (S) is plotted against the number of failure cycles (N) after implementing cyclic fatigue tests on the C- and P-series. The shapes of the fatigue test samples are flat plates as shown in the figure, and they have surface finish conditions that are identical to those listed in Table 9.22. The figure shows that the fatigue life in the C-series is superior to that in the P-series for all stress levels. The fatigue limit in the P-series is about 200 MPa, and in the C-series it is about 225 MPa, which demonstrates that the fatigue strength is also improved by the proposed grinding method. As one factor

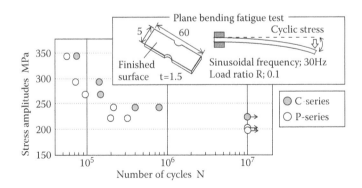

FIGURE 9.119 S-N diagram.

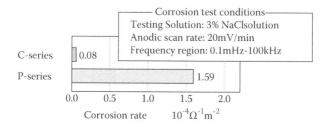

FIGURE 9.120 Results of corrosion rate test.

that contributes to the superior fatigue strength of the C-series, we consider that the oxide layer functions as a protective layer suppressing crack initiation and that compressive residual stress prevents crack propagation.

Corrosion resistance evaluation. For molds used in highly corrosive environments, corrosion resistance is one of the critical surface functionalities. Thus, electrochemical corrosion tests were performed to investigate the corrosion resistance.[56] Figure 9.120 shows the corrosion rates of the C- and P-series. The corrosion rate is significantly lower on the C-series, suggesting that the stable oxide layer produced on the work surface by the mirror-surface grinding process was thick and had a low reactivity toward corrosion.

On the other hand, in glass-forming processes, molds are exposed to high temperatures of several hundred degrees Celsius. It is necessary to evaluate the damage done on mold surfaces at high temperatures. We performed a high-temperature oxidation treatment for the C- and P-series using an electric heating furnace. Figure 9.121 shows easy macro-observations on C- and P-series after high-temperature oxidization by holding at 600°C for 10 minutes. It shows that the surface for the C-series is glossy in purple color and has less damage than that for the P-series. Figure 9.122 shows these observations in terms of surface roughness. Before high-temperature oxidization, the C- and P-series are almost equal in average roughness. However, after high-temperature oxidization, the P-series have a much higher roughness than the C-series.

Figure 9.123 shows the results of measuring elemental Fe, Cr, and O by XPS for both substrates after high-temperature oxidation treatment. In the surface structure of the C-series, spinel-type oxide ($FeCr_2O_4$) having an extremely fine and uniform

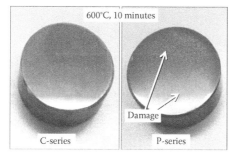

FIGURE 9.121 Macro-observations.

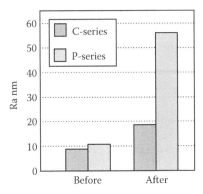

FIGURE 9.122   Surface roughness.

structure existed on the top layer. Since much oxygen was diffused into the workpiece surface by the grinding process prior to the high-temperature oxidization, $FeCr_2O_4$ could be formed instantaneously as a protective layer after high-temperature oxidization to promote stable internal oxidization with priority.[57] By contrast, only Cr_2O_3 was formed in the surface structure of the P-series. Since P-series does not form a protective external layer oxide, the oxide layer growth becomes unstable.

Adhesiveness to thin films. Precision molds are frequently coated with a hard thin film such as diamond-like carbon (DLC) or aluminum fluoride (AlF_3) for the purpose of enhancing corrosion resistance and tribological improvement, which raises issues concerning ensuring good adhesiveness in the interface between films and substrates. Thus, in this section we performed DLC coating for all series, on which scratch tests were conducted to measure the adhesive strength (see Figure 9.124). P-series had the lowest values, followed by the Al-, C-, and Si-series, in ascending order. It is known that elemental Si is utilized as an intermediate layer between substrates and DLC by

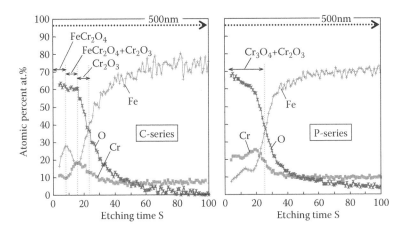

FIGURE 9.123   Results of measuring elemental Fe, Cr, and O by XPS from the surface to depth.

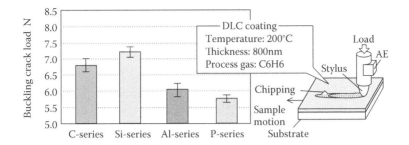

FIGURE 9.124 Results of the scratch tests.

sputtering treatment, since it has a greater chemical affinity for DLC.[58] Therefore, as a factor that imparts the Si-series with the highest adhesive strength, we consider that elemental Si, which has penetrated and diffused from abrasives into substrates, effectively acts to enhance its chemical affinity to DLC thin films.

Wettability evaluation. Wettability control of molding materials is also one of the necessary surface functionalities for molds in order to enhance mold releasability and stain resistance. Thus, we performed contact angle measurements on droplets in all series to evaluate their surface wettabilities. Figure 9.125 shows the state of droplets that have been dropped onto the workpiece surfaces and the results of the contact angle measurements. The figure shows the different shapes of each droplet. That is, the differences in abrasive elements resulted in differences in wettability on work surfaces after processing. The Al-series has the smallest contact angle.

Surface potential is one factor that affects wettability. Thus, we measured the surface potentials for all the series and show the results in Figure 9.126 together with those of the contact angle measurements. This figure shows that the contact angle tends to become smaller when increasing the surface potential. In particular, we found that only the surface potential of the Al-series has a significant positive value. As can be seen from the XPS analysis results in Figures 9.114 and 9.115,

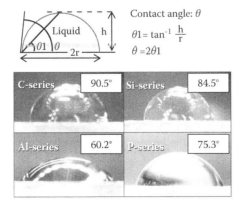

FIGURE 9.125 Results of the wettability tests.

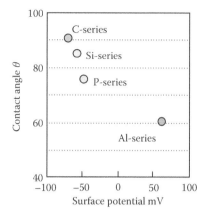

FIGURE 9.126 Surface potentials showing with contact angle.

elemental Al penetrates and diffuses into the interior of substrates in the Al-series. It is known that elemental Al is characterized by its strong positive chargeability,[59] while droplets are usually negatively charged. Therefore, we conjecture that a strong Coulomb force between the diffusion layers of elemental Al and the droplet enhance its hydrophilicity compared to the other series. These results suggest the possibility of controlling surface potential or even hydrophilicity, with respect to workpiece surfaces after processing, by changing the grinding wheels or processing conditions.

Diffusion of abrasive components into the workpiece's surface is strongly affected by the processing environment, in particular, by temperature. This section describes a basic experiment to examine the effect of the grinding fluid temperature on the diffusion of abrasive components. It involved grinding the test pieces while varying the grinding fluid temperature in the range of 25°C to 90°C, then performing elemental analysis, and evaluating the wettability. WA-series (white fused alumina) wheels and A-series (amorphous alumina) wheels were used in this experiment.[60]

Figure 9.127 shows the relationship between the peak intensity for elemental aluminum on the surface after processing and the grinding fluid temperature. The peak intensity of elemental aluminum diffusion for A-series was higher than those for WA-series at all temperatures. The result that the abrasive components of the A-series abrasive tend to diffuse more agrees with the findings given in the previous section. Figure 9.127 also shows that the peak intensity of the A-series increases sharply when the grinding fluid reaches about 50°C. This is conjectured to be due to the A-series abrasives being in the amorphous state making their behavior highly dependent on the grinding fluid temperature, as demonstrated by the mechanochemical reaction between the abrasive and the workpiece at the processing point being accelerated when the grinding fluid temperature exceeds 50°C. Figure 9.128 shows representative results of the wettability tests for both series at grinding fluid temperatures of 25°C, 50°C, and 70°C. The results for 50°C confirm that the A-series has an extremely high hydrophilicity with a contact angle of 27.8 degrees. If processing is performed at a grinding fluid temperature of 70°C or above, the processing

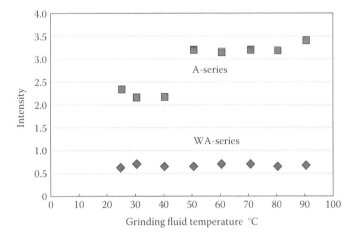

FIGURE 9.127   Relationship between the peak intensity for elemental aluminum on the surface after processing and the grinding fluid temperature.

characteristics drop in making it difficult to obtain a high-quality surface. The optimum temperature for achieving excellent processing characteristics and diffusion of abrasive components is approximately 50°C.

9.6.2   Surface Modification for Metallic Biomaterials

9.6.2.1   Surface Modification on Titanium Alloy

The properties of the finished surface. The properties of the ground surfaces by the ELID grinding method and the polished surface by SiO_2 powder were investigated by performing ultimate analysis using the energy dispersive x-ray spectroscopy (EDX). Figure 9.129 shows the results. Compared with the polished surface, the ELID ground surface exhibited a higher concentration of oxygen atoms. This suggests that the form of an oxide layer changes because of ELID grinding.[51,52] A detailed observation of the oxide layer of the ground surface was performed using the transmission

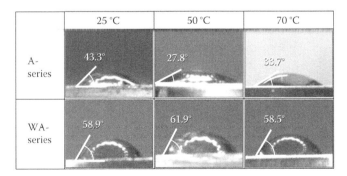

FIGURE 9.128   Results of the wettability tests for both series at grinding fluid temperatures of 25°C, 50°C, and 70°C .

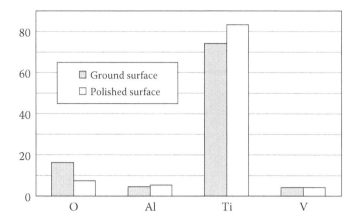

FIGURE 9.129 EDX patterns of the surface.

electron microscope (TEM). Figure 9.130 shows a cross-sectional view of the ground and polished surfaces. An oxide layer with about a 20 nm thickness in the ELID ground surface is observed. It is a lot thicker than that of the polished surface.[61,62]

We hypothesize that the observed oxygen atoms formed a stable oxide layer resulting from the electrochemical reaction that occurs during processing. Figure 9.131 shows a schematic illustration of the formation mechanism of the oxide layer. During machining, the potential electrolyte decomposes the conductive alkaline machining fluid, thereby generating hydroxide ions (OH^-). When an appropriate positive potential is applied to the workpiece, free hydroxide ions in the machining fluid are attracted to the surface, resulting in the formation of a stable oxide layer.

The possibility of controlling the formation of the oxide layer on the ELID ground surface was investigated using three types of specimens, which were ground with different voltage values of the ELID power. The oxygen atom concentration of the ground surface was measured by performing ultimate analysis by using the EDX. The result

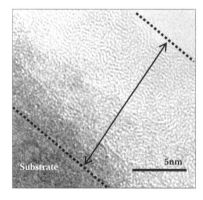
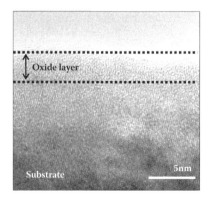

FIGURE 9.130 TEM observation of the oxide layer.

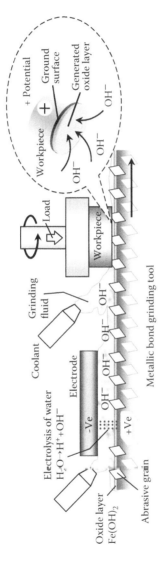

FIGURE 9.131 Schematic illustration showing the generation of oxide layer on an ELID ground surface.

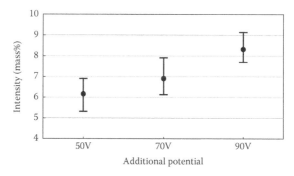

FIGURE 9.132 Change of oxygen atom concentration by the difference in additional potential.

is shown in Figure 9.132. From this figure, although some variations are accepted, as the voltage value of the ELID power was increased from 50 V to 90 V, the oxygen concentrates on the ground surface also increased. This suggests that a higher voltage applied during the process yields a thicker and more stable oxide layer.

These results support the formation mechanism of the oxide layer shown in Figure 9.131. When the voltage value of the ELID power is raised, decomposition of the conductive alkaline machining fluid is increased. Thereby, with an increase in hydroxide ions (OH^-), a stable oxide layer is formed on the surface.

The depth profile for titanium and oxygen atoms showed a similar qualitative behavior as shown in Figure 9.133 for all samples. After prolonged sputtering, the oxygen signal decreased and the titanium signal increased to their constant values in the bulk metal. The thickness of the oxide layer is defined as the depth from the surface where the curve of oxygen and titanium meets. The oxide thickness of the polished surface and the ELID ground surface was approximately 3 and 20 nm, respectively.

In addition, compared with the polished surfaces, the ELID ground surfaces exhibited a higher oxygen concentration inside the base material. This suggests that the ELID grinding method could have formed not only a thicker oxide layer but also an oxygen-diffused layer.

Figure 9.134 shows the results of the carbon concentration on the depth. Compared with the polished surface, the ELID ground surface exhibited a higher carbon concentration inside the base material. This diffusion of carbon brings about hardening of the ELID ground surface.

The exhibited carbon is a result of the electrochemical reaction that occurs during ELID grinding process. ELID grinding always performs the "dressing" of the grinding wheel, causing an electrolysis phenomenon during machining. Fresh abrasive always sticks out on the wheel surface. This abrasive reacts on the titanium surface activated by machining. In this study, since a diamond was used as an abrasive, carbon may have been diffused. To verify this, however, additional study is necessary.

Control of surface modification layer. Figure 9.135 shows a result of colored surface observation. It can be seen that a rainbow of highly attractive colors has been obtained. The proposed technique allows the brightly colored surface of red, yellow, or blue to be produced on titanium alloy. It was predicted that the thickness of the oxide layer formed on the processed surface will govern the final color tones.[63]

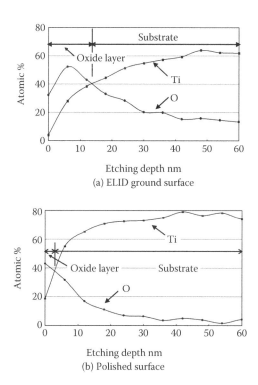

FIGURE 9.133 Results of the XPS. (a) ELID ground surface. (b) Polished surface.

To investigate this experimentally, the applied voltage and maximum current value during processing were varied, and the color tones of the processed surface obtained each time were measured. Figure 9.136 shows the colorimetric results for the *a*b colorimetric system (JIS Z 8729) obtained using a color analyzer, displayed in the a*b* coordinate system where the angle indicates the hue (the type of color), the radial

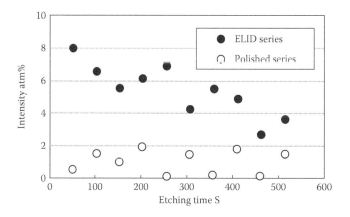

FIGURE 9.134 XPS depth profile of the carbon.

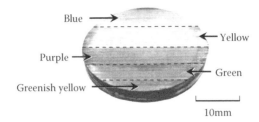

FIGURE 9.135   Result of colored surface observation.

distance indicates the chroma (the vividness of the color), and the numeric value in parentheses indicates the value of the voltage and current applied during processing.

As the applied voltage was increased, the color tones of the processed surface formed a spiral shape close to the origin, with the color changing from yellow to red, blue, green, yellow, and then red as the voltage was increased from 60 to 150 V, passing through more than a full circle. In other words, by controlling the applied voltage, all hues can be represented on the processed surface. In this example, because the shape of the circle is elliptical, there is a tendency for the chroma to be higher for a specific hue (purple).

Figure 9.137 shows the relationship between the applied voltage and the thickness of the oxide film generated on the processed surface. In this case, the film thickness was estimated based on the results of XPS analysis, as described later. The range of oxide film thickness that affects the color of the processed surface is between 100 and 250 nm, and increases in a largely linear fashion as the applied voltage increases. An oxide film of approximately 2 to 5 nm in thickness is formed on titanium upon

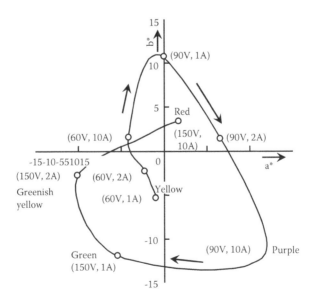

FIGURE 9.136   Colorimetry result for *a*b colorimetric system.

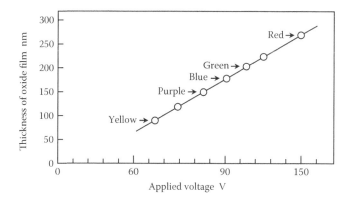

FIGURE 9.137 Relationship between applied voltage and thickness of the oxide layer.

exposure to air normally. However, the present technique was verified to allow for extremely thick oxide films of more than 200 nm in thickness to be produced.

The properties of the machined Ti-6Al-4V alloy surface were examined by chemical element analysis using an energy-dispersive x-ray (EDX) diffraction system. Figure 9.138 shows the results. It can be seen from this figure, focusing particularly on the concentration of oxygen atoms, that the peaks detected for the material modified at both the 60 V and 90 V series are sharper than those of the polished series. Furthermore, the 90 V series exhibits a stronger concentration peak than the 60 V series, consistent with the earlier results, that is, the higher the applied voltage, the thicker the oxide film that can be obtained.

The authors also analyzed the bonding statuses of the oxide films formed on various specimen surfaces. Figure 9.139 shows an example of the XPS results for analysis of titanium and oxygen in the 90 V series. The principal component of the oxide film formed on the surface appears to be TiO_2. Spectra demonstrating similar peaks were observed for both the polished series and other surface-modified samples. Figure 9.140 shows the titanium and oxygen profile in the depth direction. For the 90 V series, oxygen was detected to a depth of around 150 nm, indicating that an oxide film had formed to this depth.

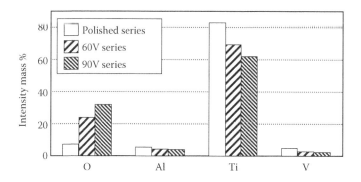

FIGURE 9.138 Analytical results using energy-dispersed x-ray diffraction system (EDX).

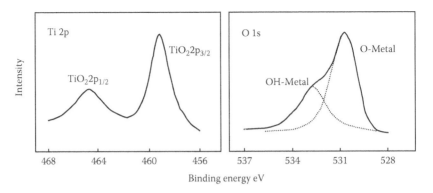

FIGURE 9.139   XPS results for analysis of titanium and oxygen in the 90 V series.

To corroborate the chemical surface analyses conducted to this point, the condition of the oxide layer formed on the surface of Ti-6Al-4V alloy processed by the proposed electrochemical grinding method was observed by transmission electron microscopy (TEM). Figure 9.141 shows a cross-sectional view of the colored surface (90 V series). An oxide (amorphous) layer of approximately 150 nm in thickness is clearly observed on top of the substrate. Based on these observations, it is thought that the oxygen detected in the surface-modified materials exists on processed surfaces in the stable oxide film state, which is consistent with an anodic oxidation reaction caused by an electrochemical reaction, as described earlier. Further experiments are to be planned to clarify the details of the formation mechanism of the oxidized layer and to determine optimum processing conditions such as electrical current and machining fluid composition.

Chemical stability of the surface. In order to investigate the chemical stability of the specimens, we measured the corrosion potential and the polarization resistance of these specimens. Figure 9.142 shows the corrosion potentials of specimens after 1 hour immersion in 0.89% NaCl solution. If the materials are immersed into an electrolyte solution in a fixed time, the surface will reach a steady state potential. This potential is defined as corrosion potential, E_{corr}. The more positive value of E_{corr}

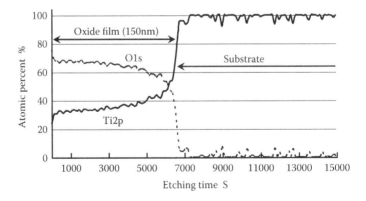

FIGURE 9.140   Titanium and oxygen profile in the depth direction of the 90 V series.

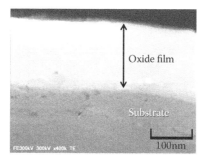

FIGURE 9.141 Cross-sectional observation by TEM of colored surface (90 V series).

indicates that it is difficult to release the metal from the specimen. It can be seen from this figure that, in comparison to the polished surface, the corrosion potential of both ground series is shifted toward the noble side. Figure 9.143 shows the results of AC impedance tests. The polarization resistance of the ELID series was higher than that of the polished series. The EG-X series clearly showed the highest polarization resistance. EG-X stands for a new compound surface modification method consisting of a new electrical grinding technique based on ELID grinding.

Corrosion potential measurements and the AC impedance test give information about the reaction between the surface and the solution in order to investigate the differences of the thickness of the layers; a detailed observation of the oxide layer was performed by using a TEM. Figure 9.144 shows a cross-sectional view of the ground and polished surfaces. The thickness of the oxide layer formed on the surface of the ELID series was about 15 nm, and in the polished series about 3 nm. In the EG-X series, an oxide layer (amorphous part) with about a 150 nm thickness was clearly observed. It is considered that a thicker oxide layer lowers the corrosion rate.

Figure 9.145 shows the results of potentiodynamic polarization measurements. In this figure, the higher the pitting potential, the higher the corrosion resistance of the workpiece. The pitting potential of the polished series was about 2.5 V, and the ELID series was 3.0 V. In the case of the EG-X series, however, no rapid increase of

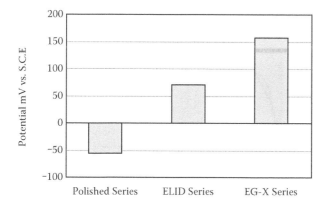

FIGURE 9.142 Comparison of the corrosion potential.

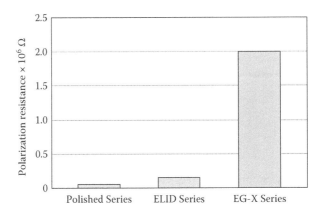

FIGURE 9.143 Comparison of the polarization resistance.

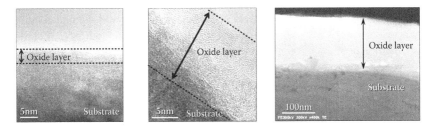

FIGURE 9.144 Cross-sectional observations by TEM.

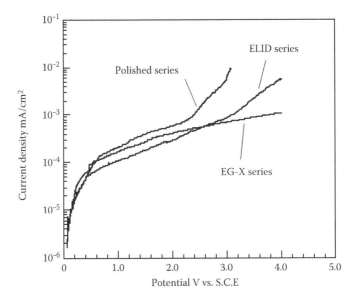

FIGURE 9.145 Cyclic polarization curves.

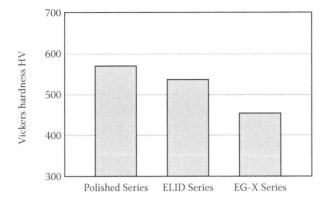

FIGURE 9.146   Results of nanoindentation test.

current density was observed. This implies that the EG-X series has higher regeneration ability as compared to those of the others.

This is because the diffused oxygen (as shown in Figure 9.134) accelerates the
regeneration of the oxide layer broken by raised potential, and prevents dissolution of
metal ions. As a result, corrosion resistance in the passive state was improved. These
results suggest that the proposed technique could improve chemical properties.

Mechanical properties of the surface. Surface hardness usually affects mechanical properties of materials. The hardness of each specimen was measured by a nano
hardness tester (NHT) with an applied load of 5 mN. Figure 9.146 shows the results.
The hardness of the EG-X series and the ELID series were higher than that of the
polished series. This may have been caused by the diffusion of carbon due to the
chemical reaction between the specimen surface and the wheel (Figure 9.135).

To investigate the effect of the surface hardening on the tribological properties,
reciprocating sliding friction tests were carried out. Figure 9.147 shows the results.
In the early stage of sliding cycles, from 200 to 500, the friction coefficients of

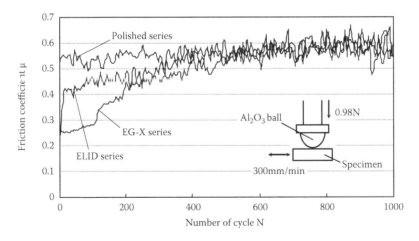

FIGURE 9.147   Results of sliding friction test.

the EG-X series and the ELID series were lower than those of the polished series. Although the friction coefficient is dependent on test conditions, both series obtained better tribological properties. At this stage, the higher hardness of the ELID ground surfaces might have been a factor contributing to the improvement in tribological properties.

Although both ELID series had low friction coefficients, their behavior was clearly different in the initial stage (from 0 to 200 cycles). In the ELID series, the value of the friction coefficient increased rapidly. On the other hand, the EG-X series maintained a low friction coefficient.

These differences may have been caused by the difference of the thickness of the oxide layer. As indicate earlier, the thickness of the oxide layer of the EG-X series clearly thicker than that of the ELID series and the polished series. Therefore, this thick and stable oxide layer generated in the EG-X series leads to maintaining a low friction coefficient in the initial stage of wear testing.

Biological properties of the surface. The effect of surface finishing on cell number is shown in Figure 9.148. In this figure, 5 days after confluence, a greater number of cells were found in the EG-X series than in the ELID series, while fewer cells were found in the polished series.

Figure 9.149 shows the cytotoxicity of extracts from all samples, expressed as an increase in lactate dehydrogenase (LDH) contents of cells. The higher the LDH contents, the lower the cytotoxicity of the workpiece. The LDH contents of the ELID series were lower than those of the polished series. The LDH contents of the EG-X series were significantly lower when compared to other series.

This is because the thick and stable oxide layer may affect corrosion resistance and biocompatibility. As indicate earlier, by comparing the corrosion resistance of each series, it was revealed that an ELID ground surface had very good corrosion resistance with very low passive current density and very high pitting potentials. The low passive current density and high pitting potentials of the ELID ground surface indicated a slower corrosion rate and subsequently a lower ion release into electrolytes or biological fluids. In addition, it is considered that keeping a low corrosion rate and a low ion release greatly accelerates cell proliferation and limits cytotoxicity.

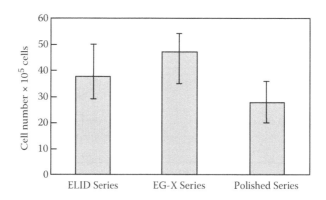

FIGURE 9.148   Cell number.

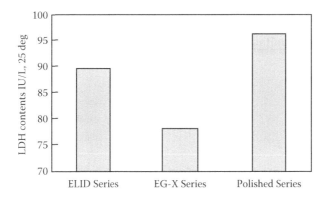

FIGURE 9.149   LDH contents.

9.6.2.2   Surface Modification on Co-Cr Alloy

Grinding characteristics. Figure 9.150 shows the results of the surface rough-ness measurements for Co-Cr alloy machined under the processing conditions as in Table 9.24. The results indicate that the finer the abrasive size of the grinding wheel, the greater the improvement in surface roughness. Also, when compared to the P-series, the G-series obtained almost the same surface roughness. In spite of employing a new grinding fluid, in which various factors such as the antiseptic and pH were adjusted, we could achieve surface roughness that corresponds to that of polishing by the ELID grinding method.[64]

Figure 9.151 shows finished surface microphotographs of both samples. In both cases, machining marks can be recognized. In particular, for the polished material, numerous scattered holes can be recognized. These holes were probably formed because loose abrasives plunge into the surface of the workpiece during the polishing process. When the material is used as a biomaterial, metal ions may leach out through these holes, generate local corrosion, and inflict serious damage to biotissues. As compared with the P-series specimen, the G-series specimen shows very few holes suggesting that the pres-ent processing method is a suitable method for the surface finishing of Co-Cr alloy.

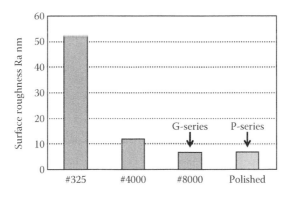

FIGURE 9.150   Surface roughness of finished Co-Cr alloy.

TABLE 9.24
Experimental Conditions

Workpiece	Co-Cr alloy	
Machine	Single-sided lapping machine	
Grinding wheel	Rougher	#325 and #4000 cast-iron bonded diamond wheels
	Finish	#8000 metal-resinoid hybrid bonded diamond wheel
Grinding fluid	Chemical solution type grinding fluid (5% dilution to water, pH 8.2)	
Grinding conditions	Wheel rotation: 100 min⁻¹; Workpiece rotation: 100 min⁻¹	
Electrical conditions	Open voltage: 90 V; Pulse timing (on/off): 2/2 μs; Pulse wave: Square	

 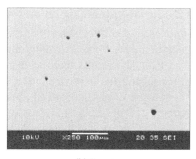

(a) G-series   (b) P-series

FIGURE 9.151 Finished surface microphotographs. (a) G-series. (b) P-series.

Additionally, test grinding was carried out to verify that this processing method could actually be used to produce a typical profile of an artificial joint. Figure 9.152 shows the tool and workpiece setup. A compact desktop-type four-axis grinding machine was used for the experiment. Finished examples are shown in Figure 9.153.

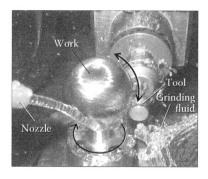

FIGURE 9.152 Tool and workpiece setup on compact desktop-type four-axis grinding machine.

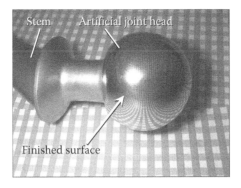

FIGURE 9.153 Ground Co-Cr alloy for artificial joint.

By use of this grinding method, an artificial joint profile with a mirror-like surface finish could be obtained in about 10 to 15 min.

Analysis of machined surface. The properties of the machined surface of Co-Cr alloy were examined by nanoindentation testing. The applied indentation load was 5 mN, and the results are shown in Figure 9.154. Whereas the hardness of the P-series was 600 HV, the G-series had a significantly higher hardness of 770 HV, confirming the superior mechanical toughness of the surface obtained using the proposed processing method.

The machined Co-Cr alloy surfaces were examined by chemical element analysis using XPS, and the results are shown in Figure 9.155. As can be seen from this figure, focusing particularly on the concentration of chromium oxide atoms in the depth direction, the peak detected for the G-series is sharper than that for the P-series. During the ELID grinding process, a high electric potential difference is generated between the conductive grinding wheel and the electrode for dressing. As a result, the electrolysis of water takes place in the grinding fluid, which has been supplied

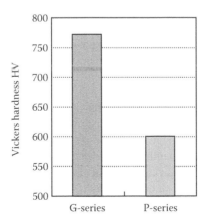

FIGURE 9.154 Results of nanoindentation test.

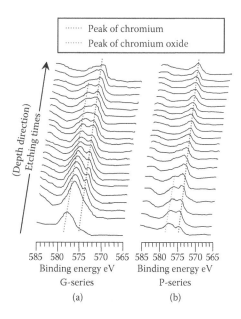

FIGURE 9.155 Detected elements profile in depth direction analyzed by using XPS. (a) G-series. (b) P-series.

to the gap, and the concentrations of hydroxide ions and dissolved oxygen are considered to increase significantly. Thus, the hydroxide ion and dissolved oxygen are presumed to have penetrated and diffused into the activated surface of the workpiece. These results suggest that the increase in surface hardness seen in Figure 9.154 is caused by the oxygen diffusion phenomenon demonstrated here.

Furthermore, the processed surface layer was analyzed for nitrogen as a detection target, and the results of the elemental analysis are shown in Figure 9.156. A nitrogen (N1s) peak is clearly observed for the G-series. On the other hand, a nitrogen peak was not observed for the P-series. According to another report, cell adhesion is improved if the surface is doped with nitrogen atoms.[59] In the following section, the cytotoxicity evaluation test is discussed by taking the above effect into account. In this experiment, only oxygen and nitrogen were analyzed as detection targets. However, recent research results clearly indicate that the ELID grinding method causes penetration and diffusion of other elements in the machined surface, such as carbon or silicon.

Evaluation of biocompatibility. The results of LDH activity measurement, which is most commonly used as a method for biocompatibility evaluation, are shown in Figure 9.157. When some damage is inflicted on a cell during the immersion experiment, LDH inside the cell membrane is released and LDH activity in the serum increases. Therefore, by measuring LDH, the degree of cell membrane damage and the degree of the cell death can be quantified. The larger this value, the higher the toxicity to the cell. As seen from Figure 9.138, the LDH activity of the G-series is less than one-fourth that of the P-series. This result clearly shows that the cytotoxicity of the G-series is extremely low.

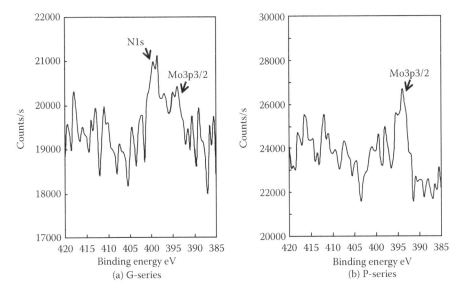

FIGURE 9.156   XPS results for analysis of nitrogen. (a) G-series. (b) P-series.

The measurement results for the number of adherent cells are shown in Figure 9.158. The number of adherent cells is the number of cells that adhere to a specimen surface after a 4-day immersion experiment. The higher the number, the better the compatibility of the surface to cells. In the figure, the G-series shows a value almost twice that of the P-series, and the cells are actively multiplying. The direct observations by SEM for adherent cells during the immersion experiment are shown in Figure 9.159. As compared with the P-series result, the G-series result showed that the cells stretch and strongly adhere to the Co-Cr alloy surface.

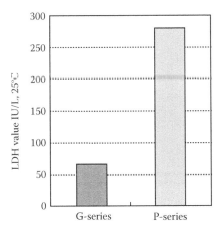

FIGURE 9.157   LDH value.

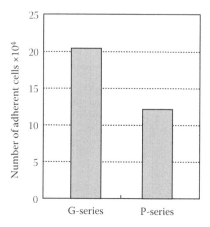

FIGURE 9.158   Number of adherent cells.

These results suggest that by processing with the grinding fluid newly developed in this present study, namely, the grinding fluid produced by removing antiseptics and setting the pH value close to that of body fluid, a surface more suitable for the multiplication of biological cells can be provided as compared with the surface finished by conventional polishing.

A number of different factors simultaneously can act to improve biocompatibility. Therefore, further experiments are planned to clarify the details of the surface modifying mechanisms and to determine optimum processing conditions, such as pH value, other components of grinding fluid, more detailed analysis of surface diffused elements, and interactions of the respective factors.

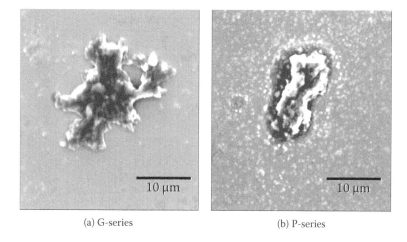

FIGURE 9.159   SEM observation of adherent cells. (a) G-series. (b) P-series.

9.7 ELID GRINDING OF SAPPHIRE: EXPERIMENTAL APPROACH

9.7.1 INTRODUCTION

Sapphire, which is a single-crystal form of α-alumina, is widely used in a range of high-technology applications, such as optics, electronics, and high temperature sensors, because of its combination of excellent optical, mechanical, physical, and chemical properties.[65–68] In many of these applications, supersurface finish and high accuracy are required. Currently, grinding is the major manufacturing process. A metal-bonded superabrasive wheel is widely used to obtain a mirror-surface finish on the advanced ceramic materials. In grinding processes, low efficiency, high abrasive wear rate, and long wheel dressing time are the main problems. Moreover, to reduce the surface roughness and subsurface damage, grinding wheels with smaller diamond grains are needed. However, such wheels are susceptible to loading and glazing, especially when fine grains are used.[69–71] Periodic dressing is required to minimize the problems in conventional grinding, which makes the grinding process less efficient. The ELID technique can continuously dress the wheel during grinding processes by electrolysis.[72–74]

In this section, the ELID technique is applied to minimize the problems in conventional grinding of sapphire. Most of the experiments were conducted to investigate the influence of wheel speed, feed rate, and depth of cut on the ground surface roughness. A mathematical model is developed for estimating the surface roughness on the basis of experimental results.

9.7.2 EXPERIMENTAL SETUP

The ELID grinding experiments were conducted on a Thompson Creep Feed Grinder equipped with Fanuc CNC and ELID system with a feed accuracy of 1 μm (Figure 9.160). A copper electrode, covering one-sixth of the perimeter of the grinding

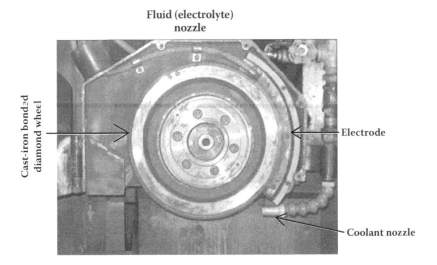

FIGURE 9.160 ELID applied to the grinding wheel.

TABLE 9.25

Experimental System Specifications

Grinding wheel	Special metal-bonded diamond wheel, #2000; Diameter: 305 mm; Width: 10 mm
	TRIM229-E (1:20), PH: 8.5~9
Electrolyte	Conductivity: 4.26 millisiemens
	On (3 μs), Off (3 μs)
Predressing	Copper
Electrode	Sapphire (diameter: 76 mm)
Workpiece measuring tool	HOMMEL-TESTER T 1000
ELID power	60 V, 9 A
Pulse width	6 μs

wheel, was used for electrolytic dressing. The metal-bonded diamond grinding wheel is connected to the positive pole through the contact of a carbon brush. The diameter of the wheel is 305 mm and the width is 10 mm. The gap between the wheel and electrode was adjusted to 0.1~0.2 mm.

A coolant, TRIM229, was diluted with distilled water to a ratio of 1:20 and used as an electrolyte and a coolant for the experiment. TRIM229 is a synthetic coolant concentrate with a refractometer factor of 1.14 and typical PH operating range of 8.5 to 9.0. The conductivity of TRIM229 at a concentration of 5% is about 4.26 millisiemens. A direct current pulse generator was used as the power supply to generate a square pulse wave for the ELID process. The surface finish was measured by the HOMMEL-TESTER T 1000. The method of measuring the surface roughness is the stylus method. The specifications and conditions of the experimental system are listed in Table 9.25.

9.7.3   CHIP THICKNESS MODEL IN ELID GRINDING

Material removal in grinding is mainly a chip formation process, in which the workpiece surface profile is generated by the grooves left by grain paths. Many attempts have been made to predict the performance of grinding processes by using the chip thickness model. The equation for estimating the maximum chip thickness or grain penetration depth in grinding[75] can be expressed as follows:

$$h_m = \left[\frac{4}{Cr} \frac{v_w}{v_s} \left(\frac{a_e}{d_e} \right)^{1/2} \right]^{1/2} \tag{9.1}$$

where h_m is the maximum undeformed chip thickness, C is the number of active grit per unit area of the wheel periphery, r is the ratio of chip width to average undeformed chip thickness, v_s is the wheel peripheral speed, v_w is the work speed, a_e is the wheel depth of cut, and d_e is the equivalent wheel diameter.

During the ELID grinding process, due to the formation of an oxide layer, the active grits on the wheel are bonded in the metal oxide matrix, which has a lower strength than the metal bond. The reduction of bond strength reduces the grit depth of cut, which is the reason for a better surface finish. In ELID grinding, the chip thickness model can be modified by adding a constant k into the existing model, then h_m can be expressed as[69]

$$h_m = k \left(\frac{4}{Cr} \frac{v_w}{v_s} \right)^{1/2} \left(\frac{a_e}{d_e} \right)^{1/2} \tag{9.2}$$

where k is the ELID constant that ranges from 0 to 1, and it is a function of dressing current, voltage, current duty ratio, and the coolant properties. The increase of the applied voltage and current duty ratio will increase the formation rate and thickness of the oxide layer on the wheel surface, which consequently results in a lower value of k and chip thickness. Therefore, in the ELID grinding, a thicker oxide layer is preferred when supersurface finish is needed. In addition, k also depends on the properties of the workpiece, such as the modulus of elasticity, because the grinding is a wheel–workpiece interaction process, where the properties of the workpiece will definitely influence the performance of the oxide layer.

The relationship between the maximum chip thickness and kinematics parameters is shown in the previous equations. It is known that the value of surface roughness is related to the chip thickness, and a good surface finish can be achieved when the chip thickness is smaller. Therefore, the equations can be used to evaluate the effect of kinematics parameters on the surface roughness.

9.7.4 EXPERIMENTAL RESULTS AND DISCUSSION

A 3^3 factorial design was used to evaluate the ELID grinding of sapphire. The effect of kinematics parameters, such as wheel speed, feed rate, and depth of cut on the surface roughness, is investigated. The depth of cut used is 1, 2, and 3 μm per pass. The RPM used is 600, 800, and 1000, which has the corresponding wheel speed 9.6 m/s, 12.8 m/s, and 16 m/s. The feed rate applied is 0.5 m/min, 1 m/min, and 2 m/min. The ELID grinding was conducted under $E_0 = 60$ V, $I_{initial} = 9A$, $\tau_{on} = \tau_{off} = 3$ μs; gap, 0.1~0.2 mm; and predressing time of 30 minutes. For each set of parameters, 5 repeated grinding tests were run, and a total of 15 roughness measurements were taken.

The results of these grinding trials show that the achieved surface roughness is as good as 50 nm when a superabrasive wheel and ELID technique are used. The effect of wheel speed, feed rate, and depth of cut on surface finish in ELID grinding of sapphire is shown in Figures 9.161, 9.162, and 9.163. It shows that wheel speed has a significant effect on the surface finish of sapphire from 9.8 m/s to 16 m/s. Feed rate also affects the surface finish of sapphire from 0.5 m/min to 2 m/min, but the effect is not as significant as wheel speed.

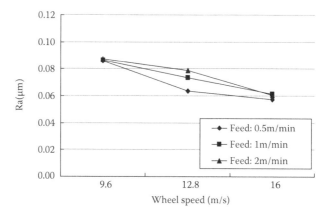

FIGURE 9.161   Effect of wheel speed on Ra.

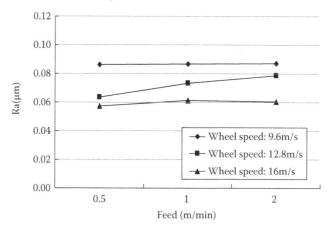

FIGURE 9.162   Effect of feed rate on Ra.

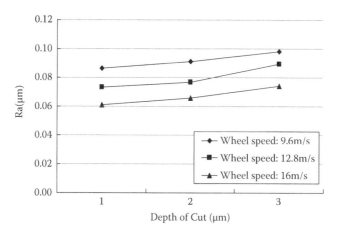

FIGURE 9.163   Effect of depth of cut on Ra.

TABLE 9.26
ANOVA for 3³ Factorial Design

Source	Sum of Squares	DOF	Mean Square	F Value	Prob > F
Model	0.058	3	0.019	318.3	
A: Wheel speed	0.044	1	0.044	726.0	<0.0001
B: Feed rate	1.318E–3	1	1.318E–3	21.60	<0.0001
C: Depth of cut	0.013	1	0.013	207.4	<0.0001
Residual	0.024	401	6.104E–5		
Lack of fit	2.358E–3	23	1.025E–4	1.75	0.0182
Pure error	0.022	378	5.851–5		

The analysis of variance (ANOVA) of the experimental results is shown in Table 9.26. The Model F-value of 318.3 implies the model is significant. There is only a 0.01% chance that a Model F-Value this large could occur due to noise. Values of Prob > F less than 0.05 indicate model terms are significant. The regression equation was determined as follows:

$$R_a =$$

$$0.11344$$

$$-4.00347\text{E-}003 \times v_s \, (\text{m/s}) \tag{9.3}$$

$$+2.89312\text{E-}003 \times v_w \, (\text{m/min})$$

$$+6.84815\text{E-}003 \times a \, (\mu\text{m})$$

From the experimental results and the regression mode, it is clear that the depth of cut, wheel speed, and feed rate affect the surface finish in the ELID grinding of sapphire, which is consistent with the theoretical chip thickness model. The increase of wheel speed, the reduction of the feed rate, or the reduction of depth of cut reduces the maximum chip thickness or grit depth of cut, and thus improves the quality of surface finish.

The estimated response surface is shown in Figures 9.164, 9.165, and 9.166.

By making use of the regression model developed, the calculated value of surface roughness and experimental results at various values of wheel speed and feed rate is shown in Figure 9.167. It can be found that there is agreement between the calculated and experimetal values.

In order to investigate the effect of electrolytic dressing parameters on the surface finish of sapphire, three additional sets of experiments were conducted by varying the electrolytic dressing current intensity. The grinding was performed at the dressing currents of 3 A, 6 A, and 9 A. The grinding parameters were wheel speed 16 m/s, feed rate 1 m/min, and depth of cut 1 μm per pass. The average surface roughness of ground surface were found to be 0.068 μm, 0.061 μm, and 0.054 μm, respectively.

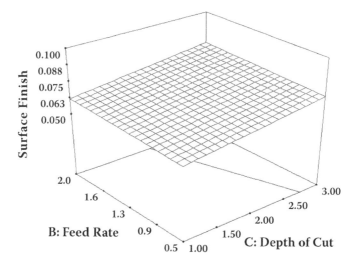

FIGURE 9.164   Estimated response surface at wheel speed 16 m/s.

These experimental results show that the increase of current intensity decreases the average surface roughness, as the increase of current intensity decreases the ELID constant k as well as the maximum chip thickness in Equation 9.2.

It is known that the ductile-mode grinding can be achieved when the maximum chip thickness or grit depth of cut is less than the critical depth of cut. For sapphire, the critical depth of cut is approximately 30 nm. The application of ELID in the grinding of sapphire provides a method to control the maximum chip thickness. By controlling the electrolytic dressing parameters, it is possible to achieve a ductile-mode grinding by using a superabrasive grinding wheel and the ELID technique, which could eliminate the requirement of finishing processes, such as lapping and polishing.

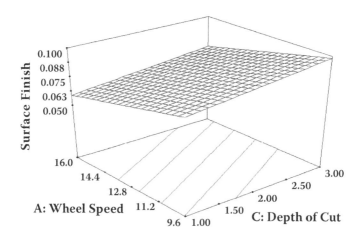

FIGURE 9.165   Estimated response surface at feed rate 1 m/min.

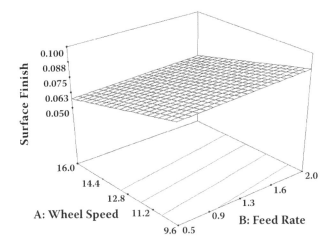

FIGURE 9.166 Estimated response surface at depth of cut 1 μm.

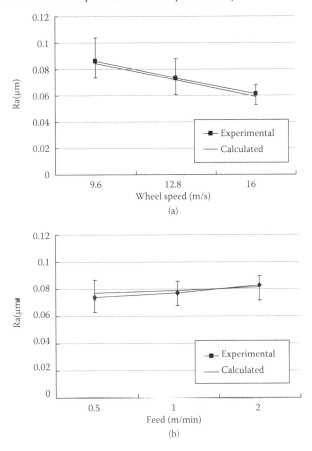

FIGURE 9.167 Comparison of calculated and experimental values of surface roughness in (a) and (b).

REFERENCES

1. N. Itoh, H. Ohmori, T. Kasai, and T. Karaki-Doy, Study of Precision Machining with Metal Resin Bond Wheel on ELID-Lap Grinding, *International Journal of Electric Machining* 3 (1998), 13–18.
2. N. Itoh, H. Ohmori, T. Kasai, and T. Karaki-Doy, Development of Metal-Resin Bonded Wheel Using Fine Metal Powder and Its Grinding Performance, *Journal of the Japan Society for Precision Engineering* 33, no. 4 (1999), 335–336.
3. N. Itoh, H. Ohmori, T. Kasai, T. Karaki-Doy, T. Iino, and H. Yamada, Finishing Characteristics of Ceramics by ELID-Lap Grinding Using Fixed Fine Diamond Grains, *International Journal of New Diamond and Frontier Carbon Technology* 10, no. 1 (2000), 1–11.
4. H. Ohashi, Power Devices: Present Status and Future Possibility, *FED Journal* 11, no. 2 (2000), 3–7.
5. K. Arai, Aspects and Prospects of SiC R&D in Japan. *IEICE Technical Report* 107, no. 110 (2007), 169–172.
6. M. Sasaki, S. Lilov, and S. Nishino, Investigation of the Hardness of Silicon Carbide, *Anuuaire de l'Université de Sofia* 97 (2004), 147–154.
7. H. Matsunami and T. Kimoto, Anisotropies in Crystal Growth of Semiconductor SiC and Its Physical Properties, *Material Integration* 17, no. 1 (2004), 3–9.
8. H. Matsunami (ed.), *Technology of Semiconductor SiC and Its Application*, Nikkan Kogyo Shinmbun Ltd., Tokyo, 10–11 (2003, 4.)
9. H. Kasuga, T. Mishima, W. Lin, Y. Watanabe, and H. Ohmori, Mirror Surface Grinding Characteristics of a Single Crystal SiC Wafer, *2007 JPSE Autumn Meeting Proceedings* (2007), 279–280.
10. H. Kasuga, H. Ohmori, W. Lin, Y. Watanabe, T. Mishima, and T. Doi, Efficient Super-Smooth Finishing Characteristics of SiC Materials through the Use of Fine-Grinding, *International Conference on Smart Manufacturing Application* (2008), 5–8.
11. N. Itoh, H. Ohmori, and T. Karaki-Doy, Effects of Metal-Resin Bonded Diamond Wheels on ELID-Lap Grinding, *Journal of Materials Processing Technology* 62 (1996), 315–320.
12. N. Itoh, H. Ohmori, T. Kasai, and T. Karaki-Doy, Study of Precision Machining with Metal Resin Bond Wheel on ELID-Lap Grinding, *International Journal of Electric Machining* 3 (1998), 13–18.
13. R. Komanduri, D. A. Lucca, and Y. Tani, Technological Advances in Fine Abrasive Processes, *CIRP Annals Manufacturing Technology* 46, no. 2 (1997), 545–596.
14. N. Itoh, H. Ohmori, and B. P. Bandopadhyay, Grinding Characteristics of a Metal-Resin Bonded Diamond Wheel on ELID Lap-Grinding, *International Journal for Manufacturing Science and Production* 1, no. 1 (1997), 9–15.
15. B. P. Bandopadhyay, Ultra-Precision and High Efficiency Grinding of Structural Ceramics by Electrolytic In-Process Dressing (ELID) Grinding, *Abrasives* (1997, April/May), 10–34.
16. E.-S. Lee, A Study on the Mirror-Like Grinding of Die Steel with Optimum In-Process Electrolytic Dressing, *Journal of Materials Processing Technology* 100 (2000), 200–208.
17. D. Dornfeld, S. Min, and Y. Takeuchi, Recent Advances in Mechanical Micromachining, *CIRP Annals Manufacturing Technology* 55, no. 2 (2006), 745–769.
18. H. Onikura, O. Ohnishi, and Y. Take, Fabrication of Micro Carbide Tools by Ultrasonic Vibration Grinding, *CIRP Annals Manufacturing Technology* 49, no. 1 (2000), 257–260.
19. Y. Takeuchi, S. Maeda, T. Kawai, and K. Sawada, Manufacture of Multiple-Focus Micro Fresnel Lenses by Means of Nonrotational Diamond Grooving, *CIRP Annals Manufacturing Technology* 51, no. 1 (2002), 343–346.

20. T. Masuzawa and M. Kimura, Surface Finishing of Tungsten Carbide Alloy, *CIRP Annals Manufacturing Technology* 40, no. 1 (1991), 199–202.

21. M. Vasile, C. Friedrich, B. Kikkeri, and R. McElhannon, Micron-Scale Machining: Tool Fabrication and Initial Results, *Journal of Precision Engineering* 2, no. 3 (1996), 180–186.

22. H. Ohmori, K. Katahira, Y. Uehara, Y. Watanabe, and W. Lin, Improvement of Mechanical Strength of Micro Tools by Controlling Surface Characteristics, *CIRP Annals Manufacturing Technology* 52, no. 1 (2003), 467–470.

23. Y. Uehara, H. Ohmori, Y. Yamagata, W. Lin, K. Kumakura, S. Morita, T. Shimizu, and T. Sasaki, Development of Small Tool by Micro Fabrication System Applying ELID Grinding Technique, *Initiatives of Precision Engineering at the Beginning of a Millennium* (2001), 491–495.

24. H. Ohmori, Y. Uehara, K. Katahira, Y. Watanabe, T. Suzuki, W. Lin, and N. Mitsuishi, Advanced Desktop Manufacturing System for Micro-Mechanical Fabrication, *Laser Metrology and Machine Performance* 7 (2005), 16–29.

25. J. S. Johnson, K. Grobsky, and D. J. Bray, Rapid Fabrication of Lightweight Silicon Carbide Mirrors, *Proceedings of SPIE* 4771 (2002), 243–253.

26. F. Pollicove and D. Golini, Deterministic Manufacturing Processes for Precision Optical Surfaces, *Key Engineering Materials* 238, no. 2 (2003), 53–58.

27. T. Bifano, Y. Yi, and W. K. Kahl, Fixed Abrasive Grinding of CVD-SiC Mirrors, *Precision Engineering* 16, no. 2 (1994), 109–116.

28. H. Suzuki, M. Hirano, M. Abe, Y. Niino, and Y. Namba, Ductile Grinding of Chemical Vapor Deposited Silicon Carbide for X-Ray Mirrors, *Journal of the JSPE* 61, no. 4 (1995), 571–575.

29. Y. Namba, H. Kobayashi, H. Suzuki, and K. Yamashita, Ultraprecision Surface Grinding of Chemical Vapor Deposited Silicon Carbide for X-Ray Mirrors Using Resinoid-Bonded Diamond Wheels, *CIRP Annals Manufacturing Technology* 48, no. 1 (1999), 277–280.

30. C. Zhang, T. Kato, W. Li, and H. Ohmori, A Comparative Study: Surface Characteristics of CVD-SiC Ground with Cast Iron Bond Diamond Wheel, *International Journal of Machine Tools and Manufacture* 40 (2000), 527–537.

31. H. Ohmori and T. Nakagawa, Mirror Surface Grinding of Silicon Wafer with Electrolytic In-Process Dressing, *CIRP Annals Manufacturing Technology* 39, no. 1 (1990), 329–332.

32. H. Ohmori and T. Nakagawa, Analysis of Mirror Surface Generation of Hard and Brittle Materials by ELID (Electrolytic In-Process Dressing) Grinding with Superfine Grain Metallic Bond Wheels, *CIRP Annals Manufacturing Technology* 44, no. 1 (1995), 287–290.

33. H. Ohmori and T. Nakagawa, Utilization of Nonlinear Conditions in Precision Grinding with ELID (Electrolytic In-Process Dressing) for Fabrication of Hard Material Components, *CIRP Annals Manufacturing Technology* 46, no. 1 (1997), 261–264.

34. N. Itoh, H. Ohmori, C. Liu, W. Lin, and T. Kasai, Super Smooth Surface Finishing of X-Ray Mirror Materials by ELID-Lap Grinding and Metal-Resin Bonded, *Proceedings of International Workshop on Extreme Optics and Sensors (International Progress on Advanced Optics and Sensors)* (2003), 83–90.

35. H. Ohmori, M. Ohmae, and K. Tanino, Ultraprecision Grinding of CVD-SiC Mirrors Using Electrolytic In-Process Dressing (ELID), *Proceedings of ELID Grinding*, 11 (1995), 245–251.

36. H. Ohmori, Y. Dai, W. Lin, et al., Force Characteristics and Deformation Behaviors of Sintered SiC during an ELID Grinding Process, *Key Engineering Materials* 238, no. 2 (2003), 65–70.

37. J. Shen, S. H. Liu, K. Yi, et al., Subsurface Damage in Optical Substrates, *OPTIK* 116, no. 6 (2005), 288–294.

38. J. A. Randi, J. C. Lambropoulos, and S. D. Jacobs, Subsurface Damage in Some Single Crystalline Optical Materials, *Applied Optics* 44, no. 12 (2005), 2241–2249.

39. S. D. Jacobs, D. Golini, Y. Hsu, B. E. Puchebner, D. Strafford, W. I. Lordonski, I. V. Prokhorov, E. Fess, D. Pietrowski, and V. W. Kordonski, Magnetorhelogical Finishing: A Deterministic Process for Optics Manufacturing, *SPIE* 2576 (1995), 372–382.

40. I. V. Prokhorov and W. I. Kordonski, New High-Precision Magnetorheological Instruments Based Method of Polishing Optics, *OSA OF&T Workshop Digest* 24 (1992), 134–136.

41. D. Golini, W. Kordonski, P. Dumas, and S. Hogan, Magnetorheological Finishing (MRF) in Commercial Precision Optics Manufacturing, *SPIE Proceedings* 3782 (1999), 80–91.

42. J. S. Johnson, K. Grobsky, and D. J. Bray, Rapid Fabrication of Lightweight Silicon Carbide Mirrors, *Proceeding of SPIE* 4771 (2002), 243–253.

43. W. Kordonski and S. Jacobs, Model of Magnetorheological Finishing, *Journal of Intelligent Material Systems and Structures* 7, no. 2 (1996), 131–137.

44. D. Golini, Precision Optics Manufacturing Using Magnetorheological Finishing (MRF), *SPIE* 3739 (1999), 78–85.

45. A. B. Shorey, S. D. Jacobs, W. I. Kordonski, et al., Experiments and Observations Regarding the Mechanisms of Glass Removal in Magnetorheological Finishing, *Applied Optics* 40, no. 1 (2001), 20–33.

46. W. Kordonski and D. Golini, Multiple Application of Magnetorheological Effect in High Precision Finishing, *Journal of Intelligent Material Systems and Structures* 13, no. 7–8 (2002), 401–404.

47. F. H. Zhang, G. W. Kang, Z. J. Qiu, et al., Magnetorheological Finishing of Glass Ceramic, *Key Engineering Materials* 257–258 (2004), 511–514.

48. H. B. Cheng, Z. J. Feng, and Y. B. Wu, Process Technology of Aspherical Mirrors Manufacturing with Magnetorheological Finishing, *Materials Science Forum* 471–472 (2004), 6–10.

49. F. H. Zhang, G. W. Kang, and Z. J. Qiu, Surface Roughness of Optical Glass under Magnetorheological Finishing, *Key Engineering Materials* 259, no. 2 (2004), 662–666.

50. V. Bagnoud, M. J. Guardalben, J. Puth, et al., High-Energy, High-Average-Power Laser with Nd: YLF Rods Corrected by Magnetorheological Finishing, *Applied Optics* 44, no. 2 (2005), 282–288.

51. E. Brinksmeier, D. A. Lucca, and A. Walter, Chemical Aspects of Machining Processes, *CIRP Annals Manufacturing Technology* 53, no. 2 (2004), 685–699.

52. C. Heinzel and N. Bleil, The Use of the Size Effect in Grinding for Work-Hardening, *CIRP Annals Manufacturing Technology* 56, no. 1 (2007), 327–330.

53. P. Rangaswamy, H. Terutung, and S. Jeelani, 1991, Effect of Grinding Conditions on the Fatigue Life of Titanium 5AL-2.5SN Alloy, *Journal of Materials Science* 26, no. 10 (1991), 2701–2706.

54. H. Ohmori and T. Nakagawa, Analysis of Mirror Surface Generation of Hard and Brittle Materials by ELID Grinding with Superfine Grain Metallic Bond Wheels, *CIRP Annals Manufacturing Technology* 44, no. 1 (1995), 287–290.

55. H. Ohmori, K. Katahira, M. Mizutani, and J. Komotori, Investigation of Substrate Finishing Conditions to Improve Adhesive Strength of DLC Films, *CIRP Annals Manufacturing Technology* 54, no. 1 (2005), 511–514.

56. American Society for Testing and Materials, *Conventions Applicable to Electrochemical Measurements in Corrosion Testing*, ASTM G3-74, 1974.

57. N. Birks, *Introduction to High Temperature Oxidation of Metals*, London, Edward Arnold, 1983.

58. A. Fayer and O. Glozman, 1995, Deposition of Continuous and Well Adhering Diamond Films on Steel, *Applied Physics Letters* 67 (1995), 2299–2304.

59. C.-J. Berg, *Wettability*, New York, Marcel Dekker, 1993.

70. H. Ohmori, K. Katahira, J. Komotori, Y. Akahane, M. Mizutani, and T. Naruse, Surface Generation of Superior Hydrophilicity for Surgical Steels by Specific Grinding Parameters, *CIRP Annals Manufacturing Technology* 58, no. 1 (2009), 503–506.

61. M. Mizutani, J. Komotori, K. Katahira, and H. Ohmori, Development of a New Integrated Machining System: Improvement of Surface Characteristics on Metallic Biomaterials with a New Electrical Grinding System, *Journal of Machine Engineering* 7, no. 1 (2007), 15–23.

62. M. Mizutani, K. Katahira, J. Komotori, and H. Ohmori, Control of Surface Modified Layer on Metallic Biomaterials by an Advanced ELID Grinding System (EG-X), *International Journal of Modern Physics B* 20, no. 25–27 (2006), 3605–3610.

63. H. Ohmori, K. Katahira, M. Mizutani, and J. Komotori, Investigation on Color-Finishing Process Conditions for Titanium Alloy Applying a New Electrical Grinding Process, *CIRP Annals Manufacturing Technology* 53, no. 1 (2004), 455–458.

64. H. Ohmori, K. Katahira, Y. Akinou, J. Komotori, and M. Mizutani, Investigation on Grinding Characteristics and Surface-Modifying Effects of Biocompatible Co-Cr Alloy, *CIRP Annals Manufacturing Technology* 55, no. 1 (2006), 597–600.

65. H. Zhu and A. Luiz, Chemical Mechanical Polishing (CMP) Anisotropy in Sapphire, *Applied Surface Science* 236 (2004), 120–130.

66. P.-K. Chandra and F. Schmid, Growth of the World's Largest Sapphire Crystals, *Journal of Crystal Growth*, 225 (2001), 572–579.

67. Y. Wang and S. Liu, Effects of Surface Treatment on Sapphire Substrates, *Journal of Crystal Growth* 274 (2005), 241–245.

68. E. Dorre and H. Hubner, *Alumina: Processing, Properties, and Applications*, New York, Springer, 1984

69. K. Fathima, K.-A. Senthil, M. Rahman, and H.-S. Lim, A Study on Wear Mechanism and Wear Reduction Strategies in Grinding Wheels Used for ELID Grinding, *Wear* 254 (2003), 1247–1255.

70. J.-H. Liu and Z.-J. Pei, ELID Grinding of Silicon Wafers: A Literature Review, *International Journal of Machine Tools and Manufacture* 47 (2007), 529–536.

71. H.-S. Lim, K. Fathima, A. Senthil Rahman, and M. Rahman, A Fundamental Study on the Mechanism of Electrolytic In-Process Dressing (ELID) Grinding, *International Journal of Machine Tools and Manufacture* 42 (2002), 935–943.

72. B. P. Bandyopadhyay and H. Qhmori, Efficient and Stable Grinding of Ceramics by Electrolytic In-Process Dressing (ELID), *Journal of Materials Processing Technology* 66 (1997), 18–24.

73. I. D. Marinescu, *Handbook of Advanced Ceramics Machining*, Boca Raton, FL, CRC Press, 2007.

74. H. Ohmori, I. Takahashi, and B. P. Bandyopadhyay, Ultra-Precision Grinding of Structural Ceramics by Electrolytic In-Process Dressing (ELID) Grinding, *Journal of Materials Processing Technology* 57 (1996), 272–277.

75. S. Malkin, *Grinding Technology: Theory and Applications of Machining with Abrasives*, Chichester, U.K., Ellis Horwood, 1989.

Index